OPTIMIZATION OF THE ELECTRON DONOR SUPPLY TO SULPHATE REDUCING BIOREACTORS TREATING INORGANIC WASTEWATER

T0315527

Luis Carlos Reyes Alvarado

Optimization of the electron donor
supply to sulphate reducing
bioreactors treating inorganic
wastewater

Joint PhD degree in Environmental Technology

UNIVERSITÉ
— PARIS-EST

Docteur de l'Université Paris-Est
Spécialité : Science et Technique de l'Environnement

Dottore di Ricerca in Tecnologie Ambientali

UNESCO-IHE
Institute for Water Education

Degree of Doctor in Environmental Technology

Thèse – Tesi di Dottorato – PhD thesis

Luis Carlos Reyes Alvarado

Optimization of the electron donor supply to sulphate reducing bioreactors treating inorganic wastewater

Defended on December 16th, 2016

In front of the PhD committee

Dr. Artin Hatzikioseyian	Reviewer
Prof. Dr. Erkan Sahinkaya	Reviewer
Prof. Dr. Ir. Piet Lens	Promotor
Dr. Eldon R. Rene	Co-promotor
Prof. Dr. Giovanni Esposito	Co-promotor
Prof. Dr. Michel Madon	Co-promotor
Hab. Dr. Eric D. van Hullebusch	Examiner

ERASMUS MUNDUS

Erasmus Joint Doctorate Programme in Environmental Technology for Contaminated Solids, Soils and Sediments (ETeCoS^3)

Thesis commettee

Promotor
Prof. Dr. Ir. Piet N. L. Lens
Professor of Environmental Biotechnology
UNESCO-IHE
Delft, The Netherlands

Co-promotors
Dr. Eldon R. Rene
Senior Lecturer in Resource Recovery Technology
UNESCO-IHE
Delft, The Netherlands

Prof. Dr. Giovanni Esposito
Professor in Environmental Engineering
University of Cassino and Southern Lazio
Cassino, Italy

Prof. Dr. Michel Madon
Professor in Biogeochemistry
University of Paris-Est
Marne-la-Vallée, France

Other members
Dr. Artin Hatzikioseyian
School of Mining and Metallurgical Engineering
National Technical University of Athens (NTUA), Greece

Prof. Dr. Erkan Sahinkaya
Professor of Bioengineering
Medeniyet Üniversitesi
Goztepe, Istanbul, Turkey

Hab. Dr. Eric D. van Hullebusch
Hab. Associate Professor in Biogeochemistry
University of Paris-Est
Marne-la-Vallée, France

This research was conducted under the auspices of the Erasmus Mundus Joint Doctorate in Environmental Technologies for Contaminated Solids, Soils, and Sediments (ETeCoS^3) and the Graduate School for Socio-Economic and Natural Sciences of the Environment (SENSE).

CRC Press/Balkema is an imprint of the Taylor & Francis Group, an informa business

Published by:
CRC Press/Balkema
Schipholweg 107C, 2316 XC, Leiden, the Netherlands
Pub.NL@taylorandfrancis.com
www.crcpress.com – www.taylorandfrancis.com
ISBN: 978-1-138-34331-3

Table of contents

List of figures

List of figures

List of tables

List of tables

Acknowledgments

I would like to thank the European Comision for providing financial support through the Erasmus Mundus Joint Doctorate Programme ETeCoS[3] [Environmental Technologies for Contaminated Solids, Soils and Sediments, grant agreement FPA no. 2010-0009].

I would like to acknowledge the board members of the Erasmus Mundus Joint Doctorate Programme Environmental Technologies for Contaminated Solids, Soils and Sediments (ETeCoS[3]): Hab. Dr. Eric D. van Hullebusch, Université Paris Est, UPE (France); Prof. Dr. Giovanni Esposito, Università degli studi di Cassino e del Lazio Meridionale, UNICLAM (Italy) and Prof. Dr. Piet N. L. Lens, UNESCO-IHE (The Netherlands) for giving me the opportunity to participate and for being my examiner, co-promotor and promotor, respectively, in this joint Erasmus Mundus doctorate programme.

I would like to thank my main hosting institution during the PhD and the staff members at UNESCO-IHE Institute for Water Education.

I would like to thank the institutes and the staff memebers that hosted me during the PhD at: Università degli studi di Cassino e del Lazio Meridionale, UNICLAM (Italy); Laboratory of Environmental Biotechnology (LADISER Gestion y Control Ambiental) at the Universidad Veracruzana, Veracruz, Mexico; INRA, Laboratoire de Biotechnologie de l'Environnement, Narbonne, France. I would specially like to thank to Stefano Papirio, Elena Rustrian, Eric Houbron, Gaëlle Santa-Catalina, Frédéric Habouzit, Renau Escudie and Nicolas Bernet for all the scientific discussions and guidance through the research.

I would like to thank Eldon R. Rene, co-promotor of this PhD thesis, for all the scientific support, advice and discussion along these years.

I would like to express my deepest gratitude and love to my parents Oscar Reyes Herrera and Eloisa Alvarado Cruz, thank you for everything.

Abstract

Many industrial wastewaters, particularly from the mining, fermentation and food processing industry contain high sulphate (SO_4^{2-}) concentrations. SO_4^{2-} reduction (SR) is a serious environmental problem under non-controlled conditions, which can result in the release of toxic sulphide to the environment. Typical characteristics of SO_4^{2-}-rich wastewater are 0.4-20.8 g $SO_4^{2-}.L^{-1}$, low pH, high oxidative potential, low or negligible chemical oxygen demand (COD) and high heavy metals concentrations (acid mine drainage), that can dramatically damage the flora and fauna of water bodies. The aim of this PhD is to study the effect of electron donor supply on the biological SR process by sulphate reducing bacteria in bioreactors. The biological SR process was studied using carbohydrate based polymers (CBP) and lignocelulosic biowaste (L) as slow release electron donors (CBP-SRED and L-SRED, respectively) in batch bioreactors and continuously operated inverse fluidized bed bioreactors (IFBB). IFBB were vigorously tested for SR under high rate and transient (feast-famine) feeding conditions. In another bioreactor configuration, a sequencing batch bioreactor, the effect of the initial SO_4^{2-} concentration on the reactor start-up was investigated. Besides, the effect of the initial concentration of electron donor, NH_4^+, and SO_4^{2-} were evaluated in batch bioreactors as well.

The robustness and resilience of SR was demonstrated in IFBB using lactate as the electron donor wherein the SO_4^{2-} removal efficiency (SRE) was comparable in the feast period (67 ± 15%) of IFBB2 to steady feeding conditions (71 ± 4%) in the same IFBB2 and to IFBB1, the control reactor (61 ± 15%). From artificial neural network modeling and sensitivity analysis of data of IFBB2 operation, it was envisaged that the influent SO_4^{2-} concentrations affected the COD removal efficiency, the sulphide production and pH changes. In another IFBB3 at a COD:SO_4^{2-} ratio of 2.3, SR under high rate conditions (HRT = 0.125 d) was 4,866 mg SO_4^{2-}. L^{-1} d^{-1} and a SRE of 79% was achieved. Besides, the Grau second order and the Stover-Kincannon substrate removal models fitted the high rate reactor performance with $r^2 > 0.96$. The COD:SO_4^{2-} ratio was the major factor affecting the SR.

In batch bioreactors, using filter paper as CBP-SRED, SR was carried out at > 98% SRE. Using scourer as L-SRED, a 95% SRE was achieved. However, when the scourer was used as the L-SRED carrier material in an IFBB4, the SR showed 38 (± 14) % SRE between 10-33 d of operation. The SR was limited by the hydrolysis-fermentation rate and, therefore, the complexity of the SRED. Concerning sequencing batch bioreactor operation, the SR process was affected by the initial SO_4^{2-} concentration (2.5 g $SO_4^{2-}.L^{-1}$) during the start-up. The sequencing batch bioreactor performing at low SRE (< 70%) lead to propionate accumulation,

Abstract

however, acetate was the major end product when SRE was > 90%. In batch bioreactors, the NH_4^+ feast or famine conditions affected the SR rates with only 4% and the electron donor uptake was 16.6% greater under NH_4^+ feast conditions. The electron donor utilization via the SR process improved simultaneously to the decreasing initial electron donor concentrations.

This PhD research demonstrated that the SR process is robust to transient and high rate feeding conditions. Moreover, SR was mainly affected by the approach how electron donor is supplied, e.g. as SRED or as easy available electron donor, independently of the $COD:SO_4^{2-}$ ratio. Besides, the electron donor and SO_4^{2-} utilization were affected by the lack or presence of nutrients like NH_4^+.

Keywords: sulphate reducing bacteria (SRB); carbohydrate based polymers; lignocellulose; slow release electron donor (SRED); inverse fluidized bed bioreactors (IFBB); artificial neural networks; sequencing bacth bioreactors (SBR); fest-famine or transient feeding conditions

Resumé

Une grande quantité des eaux usées, particulièrement celles provenant des l'industries minière, des fermentations *et* alimentaires, contiennent des concentrations élevées en sulfate (SO_4^{2-}). La réduction du SO_4^{2-} représente une problématique sérieuse au niveau environnemental en conditions non contrôlées. Cette problématique peut générer une libération de sulfure toxique dans l'environnement. Les caractéristiques typiques des ces eaux usées riches en SO_4^{2-} sont 0.4-20.8 g $SO_4^{2-}.L^{-1}$, un faible pH, un fort potentiel oxydatif, une faible ou négligeable demande chimique en oxygène (DCO) et de fortes concentrations de métaux lourds (drainage minier acide) qui peuvent drastiquement endommager la flore et la faune des masses d'eau. L'objectif de cette thèse est d'étudier l'effet du donneur d'électrons sur le processus biologique d'élimination de sulfates (RS) par des bactéries sulfate réductrices en bioréacteurs. Le processus biologique RS a été étudié à l'aide de polymères à base d'hydrate de carbone (PBHC) et de biodéchets lignocellulosiques (L) comme donneurs d'électrons a libération lente (PBCH-DLLE et L-DLLE, respectivement) dans des bioréacteurs discontinus et continu de type lit fluidisé inverse (ILF). Les ILF ont été rigoureusement testé pour la RS sous des conditions d'alimentation forte et transitoires (alimentation-carence de substrat). Dans un autre réacteur en mode batch, l'effet de la concentration initiale en SO_4^{2-} sur le démarrage a été étudié. Par ailleurs, l'effet de la concentration initiale en donneur d'électrons, (NH_4^+ et SO_4^{2-}) a également été évaluée en réacteur batch.

La robustesse et la fiabilité de la RS ont été démontrées dans les ILF en utilisant le lactate comme source de donneur d'électrons, pour lesquels l'efficacité d'élimination du SO_4^{2-} (EES) était comparable sur la période de démarrage de ILF2 (67 ± 15%), et sur les périodes stables (71 ± 4%) pour les ILF2 et ILF1 et le réacteur de contrôle (61 ± 15%). De la modélisation des réseaux de neurones artificiels et de l'analyse de sensibilité des données de fonctionnement de ILF2, il était prévu que les concentrations de SO_4^{2-} de l'influents affecterai le rendement d'élimination de la DCO, la production de sulfure et des changements de pH. Dans un autre ILF3 à un rapport DCO:SO_4^{2-} de 2.3, la RS en condition de forte charge (CTE = 0.125 d) présente une valeur 4,866 mg SO_4^{2-} L^{-1} d^{-1} avec une EES de 79%. Par ailleurs, le second ordre de Grau et les modèles de consommation de substrat de Stover-Kincannon définissent la performance du réacteur à forte charge avec un $r^2 > 0.96$. Le rapport DCO:SO_4^{2-} étant le principal facteur d'influence de la RS.

Dans les bioréacteurs batch, en utilisant du papier filtre comme source de PBCH-DLLE, la RS a a atteint une EES > 98%. Avec l'utilisation d'éponge naturelle en tant que L-DLLE, un EES

Resumé

de 95% a été observée. Cependant, lorsque cette dernière est utilisée comme matériau support de L-DLLE dans un ILF4, le RS atteint une valeur de 38% ($\pm$ 14) de EES entre les jours 10-33 de fonctionnement. Dans ce cas, la SR était limitée par le taux d'hydrolyse-fermentation du support et en conséquence para la complexité de la DLLE. En ce qui concerne l'opération du bioréacteur en mode discontinu, le processus de RS est affecté par la concentration initiale en SO_4^{2-} (2.5 g $SO_4^{2-}.L^{-1}$) sur la période de démarrage. Lorsque le bioréacteur séquentiel presente une faible EES (< 70%) il génère une accumulation de propionate. Cependant, l'acétate était le principal produit final lorsque la EES était > 90%. Dans les bioréacteurs discontinus, les conditions d'alimentation de NH_4^+ ont affecté positivement de 4% les taux de RS et de 16% celui d'absorption du donneur d'électrons pendant le régime d'alimentation en NH_4^+. L'utilisation du donneur d'électrons pour le processus RS s'est améliorée de façon inversement proportionnel a la diminution de la concentration initiale en donneur d'électrons.

Cette recherche de doctorat a démontré que le processus de RS est robuste aussi bien en condition transitoire que pour des fortes charges. De plus, la RS est sensible au mode d'apport du donneur d'électrons, qu'il soit sous forme de SRED ou bien un donneur facilement assimilable et ce indépendamment du rapport $DCO:SO_4^{2-}$. Par ailleurs, l'utilisation du donneur d'électrons et du SO_4^{2-} sont sensible a l'absence ou la présence de nutriments tel que le NH_4^+.

Mots-clés: bactéries sulfate réductrices (BSR); polymères à base d'hydrates de carbone; lignocellulose; donneur d'électrons à libération lente (DLLE); lit fluidisé inverse (ILF), bioréacteurs batch (BBS), alimentation-Carence ou conditions d'alimentation transitoires.

Samenvatting

Industrieel afvalwater, vooral afvalwater afkomstig uit de mijnbouw, fermentatie en de voedsel verwerkings industrie, bevat vaak hoge concentraties sulfaat (SO_4^{2-}). Sulfaatreductie (SR) is een ernstig milieu probleem als het onder ongecontroleerde omstandigheden plaatsvindt, want het kan resulteren in het vrijkomen van giftig sulfide in het milieu. Kenmerkende karakteristieken van sulfaatrijk afvalwater zijn 0.4-20.8 g $SO_4^{2-}.L^{-1}$, een lage pH waarde, een hoge redoxpotentiaal, een lage- of verwaarloosbare chemische zuurstofvraag (COD) en hoge concentraties zware metalen (acid mine drainage), wat dramatische gevolgen kan hebben voor de flora en fauna van watergebieden. Dit onderzoek was gericht op het bestuderen van de toevoeging van elektrondonoren, en welk effect dit heeft op biologische sulfaatreductie in bioreactoren. Er is getracht het biologische sulfaatreductie proces met behulp van slow release elektron donoren te laten verlopen, met onder andere op koolhydraat lijkende polymeren en lignocellulose-rijk organisch afval (respectievelijk CBP-SRED en L-SRED). Deze elektron donoren werden getest in batch bioreactoren en continu opererende inverse gefluïdizeerde bed bioreactoren (IFBB). De IFBB werd intensief getest op sulfaatreductie onder afwisselend rijke en arme voedings¬condities (feast-famine). Ook het effect van de initiële SO42- concentratie op het opstarten van een bioreactor werd onderzocht in een andere bioreactoropstelling (een sequencing batch reactor). Daarnaast werd ook het effect van de initiële concentratie van elektron donor, NH_4^+ en SO_4^{2-} bestudeerd in batch bioreactoren.

De robuustheid en veerkracht van sulfaatreductie werden aangetoond in de IFBB door melkzuur te gebruiken als elektron donor, waarbij de sulfaat verwijderingsefficiëntie (SRE) gedurende een feest periode ($67 \pm 15\%$) in IFBB2 dezelfde was als in vastenomstandigheden ($71 \pm 4\%$) in dezelfde IFBB2 en de controle reactor IFBB1 ($61 \pm 15\%$). Neuraal netwerk-modellering en een gevoeligheids analyse van de data van de IFBB2 gaven aan dat de influent SO_4^{2-} concentraties de COD verwijderings efficiëntie, sulfide productie en pH veranderingen beïnvloedden. In een andere IFBB3 werd een sulfaatreductie snelheid van 4.8 g $SO_4^{2-}.L^{-1}$ d^{-1} en een SRE van 79% bereikt bij een COD:SO_4^{2-} ratio van 2:3 en onder high-rate omstandigheden (HRT = 0.125 d),. De data werden gefit aan de Grau tweede orde en de Stover-Kincannon substraat verwijderings-modellen, die een fit gaven met de high-rate reactor met $r^2 > 0.96$. De COD:SO_4^{2-} ratio was de belangrijkste factor die de sulfaatreductie beïnvloedde.

In batch bioreactoren, waarbij filter papier als CBP-SRED gebruikt werd, werd sulfaatreductie efficiëntie (SRE) van > 98% behaald. Met schuurspons als L-SRED werd een SRE van 95% behaald. Een sulfaatreductie efficiëntie van 38 ($\pm$ 14)% SRE tussen dag 10 en 33 van de run,

Samenvatting

wanneer de schuurspons werd gebruikt als L-SRED-drager materiaal in IFBB4. De sulfaatreductie werd verminderd door afbraak van de schuurspons door hydrolyse en fermentatie. Bij het draaien van sequencing batch bioreactoren werd duidelijk dat het sulfaatreductie proces tijdens de opstart wordt beïnvloed door de initiële SO_4^{2-} concentratie (2.5 g $SO_4^{2-}.L^{-1}$). Een sequencing batch bioreactor die bij een lage SRE (< 70%) draaide gaf een ophoping van propionaat. Bij een SRE van > 90% was acetaat echter het belangrijkste eindproduct. In batch bioreactoren beïnvloedde de afwisseling van feest/vasten omstandigheden van NH_4^+ de sulfaatreductie snelheden met slechts 4%, terwijl 16.6% meer elektron donor werd gebruikt onder NH_4^+ feest omstandigheden. Bij afnemende initiële elektron donor concentraties werd het gebruik van de elektron donor voor het sulfaatreductie proces verbeterd.

Dit PhD onderzoek toonde aan dat het sulfaatreductie proces robuust is bij overgangs-condities en condities met een hoge voedingssnelheid. Bovendien werd sulfaatreductie voornamelijk beïnvloed door de manier waarop elektron donor wordt toegevoegd, bijvoorbeeld als SRED, of als gemakkelijk beschikbare elektron donor, en stond het los van de $COD:SO_4^{2-}$ ratio. Bovendien werden elektron donor en SO_4^{2-} gebruik beïnvloed door de aan- of afwezigheid van voedingsstoffen zoals NH_4^+.

Trefwoorden: sulfaat reducerende bacteriën (SRB); koolhydraat-gebaseerde polymeren; lignocellulose; slow release elektron donor (SRED); omgekeerde gefluidiseerde bioreaktoren (IFBB); sequencing batch bioreactoren (SBR); feast-famine voedingscondities.

Sommario

Numerose acque reflue, in particolare quelle provenienti da attività minerarie e dall'industria alimentare, contengono alte concentrazioni di solfato (SO_4^{2-}). La solfato riduzione (SR) può rappresentare un serio problema ambientale qualora avvenga in condizioni non controllate, e può risultare nel rilascio di quantitativi tossici di solfuro nell'ambiente. Tipicamente, reflui ricchi di SO_4^{2-} contengono tra gli 0.4 e i 20.8 g SO_4^{2-} L^{-1}, basso pH, alto potenziale d'ossidazione, bassa o trascurabile domanda chimica d'ossigeno (COD) e alta concentrazione di metalli (drenaggio acido di miniera), che possono danneggiare drammaticamente la flora e la fauna nei corsi d'acqua. Lo scopo di questo dottorato è quello di studiare l'effetto dell'aggiunta di donatori di elettroni sul processo biologico di SR, mediante batteri solfato riduttori all'interno di bioreattori. Il processo biologico SR è stato studiato usando polimeri a base di carboidrati (CBP) e rifiuti lignocellulosici (L) come donatori di elettroni a lento rilascio (CBP-SRED e L-RED, rispettivamente) in bioreattori batch e in bioreattori a letti fluidizzati inversi operati in continuo (IFBB). I reattori IFBB sono stati testati per la SR ad in condizioni di alimentazione ad alto carico e intermittenti (abbondanza-carenza). In un'altra configurazione reattoristica, ovvero in un bioreattore batch sequenziale, è stato studiato l'effetto della concentrazione iniziale di SO_4^{2-} nella fase di start-up del reattore. Inoltre, l'effetto della concentrazione iniziale dei donatori di elettroni, NH_4^+ e SO_4^{2-}, è stato valutato in bioreattori batch.

La robustezza e la resistenza della SR è stata dimostrata nell'IFBB usando lattato come donatore di elettroni. Nel reattore, l'efficienza di rimozione di SO_4^{2-} (SRE) durante il periodo di abbondanza (67 ± 15%) dell'IFBB2 è stata ritenuta comparabile alle condizioni di alimentazioni stazionarie nello stesso IFBB2 (71 ± 4%) e nel reattore di controllo IFBB1 (61 ± 15%). Dalla modellazione di una rete neurale artificiale, e dall'analisi di sensitività dei dati sull'operatività dell'IFBB2, è stato rilevato che le concentrazioni di SO_4^{2-} nell'affluente avessero ripercussioni sull'efficienza di rimozione del COD, sulla produzione di solfuro e sui cambi nel pH. In un altro IFBB3, con un rapporto COD:SO_4^{2-} di 2.3, l'SR con tempo di ritenzione idraulico di 0.125 giorni è stato di 4.8 g SO_4^{2-}. L^{-1} d^{-1}, consentendo di raggiungere un SRE del 79%.

In bioreattori batch, utilizzando carta da filtro come CBP-SRED, la SR è stata effettuata ad un SRE superiore al 98%. Utilizzando spugne vegetali come L-SRE, è stata ottenuta una SRE del 95%. Tuttavia, quando la spugna è stata utilizzata come carrier per l'L-SRED nel reattore

Sommario

IFBB4, l'SR ha raggiunto un 38 ($\pm$ 14)% di SRE tra il decimo e il trentatreesimo giorno di operatività.

Con la seguente ricerca è stato dimostrato che l'SR rappresenta un processo robusto in condizioni di alimentazione intermittenti e di alto carico. Inoltre, l'SR è stata principalmente condizionata dalla modalità con cui è stato fornito il donatore di elettroni, ad esempio come SRED o come donatore facilmente disponibile , indipendentemente dal rapporto COD:SO_4^{2-}. L'utilizzo del donatore di elettroni e dell'SO_4^{2-} sono state condizionate dall'assenza o dalla presenza di nutrienti come NH_4^+.

Parole chiave: batteri solfato riduttori (SRB); polimeri a base di carboidrati; materiali lignocellulosici; donatore di elettroni a lento rilascio (SRED); bioreattori a letti fluidizzanti inversi (IFBB); reattori batch sequenziali (SBR); abbondanza-carenza o condizioni di alimentazione intermittenti.

Chapter 1
Introduction

1.1 Background

Anthropogenic activities such as mining, fermentation and food processing industry produce industrial wastewaters containing high sulphate (SO_4^{2-}) concentrations. For instance, mineral extraction, coal and processes related to mining are responsible of the generation of acid drainage, wherein sulphate is produced by iron sulphide (pyrite or FeS_2) oxidation. The main factors improving iron sulphide oxidation mediated by sulphur oxidizing bacteria are water, oxygen and a small chemical oxygen demand (COD) concentration from the surrounding mining area (Borrego *et al.*, 2012). Most sulphate rich, 0.4-20.8 g $SO_4^{2-}.L^{-1}$, wastewaters are characterized by a very low pH (2-3), lack or low COD concentration and highly oxidative conditions, *e.g.* characteristics of acid mine drainage are resumed in Table 1-1. With such characteristics, sulphate rich wastewater can dissolve other minerals (Monterroso and Macías, 1998). Moreover, the sulphate rich wastewater characteristics and mineral bioavailability might affect water reservoirs (Sarmiento *et al.*, 2009), groundwater and soils (Mapanda *et al.*, 2007).

In nature, the sulphate content in wastewaters can be reduced to sulphide by sulphate reducing bacteria (SRB), these microorganisms are capable of sulphate reduction throughout a dissimilatory pathway under anaerobic condition (Muyzer and Stams, 2008). The reduction of the sulphate content in industrial wastewater under non controlled conditions has affected rivers, like Rio Tinto and Odiel in Spain (Nieto *et al.*, 2007) and the Pearl River in subtropical China (Lin *et al.*, 2007), for example. On the other hand, sulphate reduction under controlled conditions (*e.g.* bioreactors) drives to sulphide production and further utilization of elemental sulphur produced from the sulphide (Hulshoff Pol *et al.*, 1998). Even though there are sulphide emissions in nature due to the biogeochemical sulphur cycle (Reese *et al.*, 2008), under any circumstance, hydrogen sulphide antropogenically produced must not be wasted to the environment as this acid is poisonous and toxic to humans and animals (Kage *et al.*, 2004) apart from causing serious and expensive problems like corrosion in pipes and metal structures (Vollertsen *et al.*, 2008).

Therefore, the main milestone of sulphate rich wastewater treatment is to focus on the removal of sulphate, utilization of the produced sulphide and to recover other resources from the wastewater. Different strategies have been used on treating sulphate rich wastewater, for instance to neutralize and precipitate metals using calcium carbonate from other processes (Mulopo and Radebe, 2012; Strosnider and Nairn, 2010), also several electron donors have been used for sulphate reduction, for example those like municipal wastewater, lactate, cheese whey, hydrogen and carbon dioxide (Bai *et al.*, 2013; Deng and Lin, 2013; Foucher *et al.*, 2001;

Jiménez-Rodríguez *et al.*, 2009; Yi *et al.*, 2007). In combination to different reactor configurations, batch, sequencing batch and continuous bioreactor have also been tested for the same purpose (Bai *et al.*, 2013; Deng and Lin, 2013; Foucher *et al.*, 2001; Jiménez-Rodríguez *et al.*, 2009; Yi *et al.*, 2007).

Table 1-1. Some physico-chemical characteristics of acid mine drainage (AMD)

Country	pH	SO_4^{-2} (mg.L^{-1})	EC** (mS.cm^{-1})	Eh*** (mV)	Metals	COD (mg.L^{-1})	Author
Spain	2.23-7.99	424-7404	0.82-6.51	143-754	F, Ca, Mg, Na, K, Si, Al, Fe, Mn, Ni, Co, Zn, Cu, Pb and Cd.		(Monterroso and Macías, 1998)
Spain	< 3		17.31		Sc, Y, U, and La, Ce, Pr, Nd, Sm, Eu, Gd, Tb, Dy, Ho, Er, Yb, Lu, REE*		(Borrego *et al.*, 2012)
Spain	2.3	8500			Fe, Cu, Zn, Al, Mn, Ni, Cd and Cr		(Jiménez-Rodríguez *et al.*, 2009)
Portugal	≈ 2	3100			K, Ca, Ti, V, Cr, Mn, Fe, Ni, Cu, Zn, As, Se, Rb, Sr, Cd, Sn, Sb, Ba and Pb		(Martins *et al.*, 2009)
USA	4.2±0.9	1846±594	21980±4870		Fe, Ca, Mg, Mn, Al and Na	41±49	(Deng and Lin, 2013)
India	2.3-7.6	176–3615	0.785-6.760		Na, K, Ca, Mg, Cr, Ni, Zn, Mn, Fe, Al, Cd, Pb, Cu and Co		(Equeenuddin *et al.*, 2010)
China	2.75	20800			Cu, Fe and Mn	< 100	(Bai *et al.*, 2013)
France	2.55	5800			Fe, Zn, Cu, Al, Mn, Co, Ni and Pb		(Foucher *et al.*, 2001)
Zimbabwe	2.1-2.7	16300-19000	11.4-14.1		As, Ni and Fe		(Mapanda *et al.*, 2007)

Note: *Rare Earth Elements (lanthanide series), **EC = electrical conductivity, ***Eh = redox potential

This PhD thesis aims to optimize the electron donor supply to sulphate reducing bioreactors treating such inorganic wastewaters, *i.e.* rich in sulphate concentration and lack of COD. The particular objectives are: i) to determine the robustness and resilience of biological sulphate reduction (BSR) under transient feeding conditions; ii) to determine the robustness and resilience of BSR under high rate feeding conditions; iii) to elucidate possible drawbacks influenced by the initial sulphate concentration on the start up of sequencing batch bioreactors for sulphate reduction; iv) to elucidate the influence of the initial electron donor, NH_4^+ and sulphate concentration on the sulphate reduction; v) to study the BSR using carbohydrate based polymers as slow release electron donors and vi) to study the BSR using lignocellulosic polymers as slow release electron donors.

1.2 The PhD thesis structure

The structure of this PhD thesis "Optimization of the electron donor supply to sulphate reducing bioreactors treating inorganic wastewater" is described in Figure 1-1. This PhD thesis is composed of 10 chapters, wherein all the particular objectives were included.

The PhD thesis approached the BSR within **Chapter 1**; the first chapter introduces the field of the research, inorganic wastewaters rich in sulphate and lack of COD concentration, describes the general and particular objectives and describes the thesis structure. **Chapter 2** reviewed the literature for the electron donor utilization, different reactor configurations, and transient conditions in sulphate reduction performance. Later, the research was divided based in the two main types of electron donors tested in this research. The easy available electron donors refers to those that SRB can use directly and the slow release electron donors to those that have to be hydrolysed and fermented before SRB can use them (Figure 1-1).

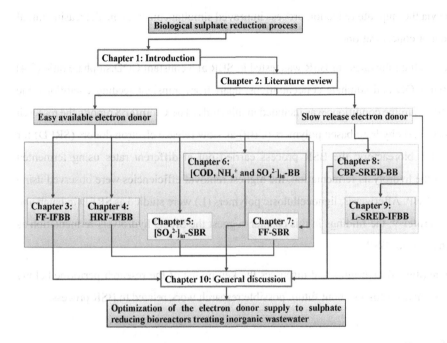

Figure 1-1. Structure of the PhD thesis
Wherein: FF-IFBB: feast-famine in an inverse fluidized bed bioreactor; HRF-IFBB: high rate feeding conditions in an inverse fluidized bed bioreactor; BB: batch bioreactor; SBR: sequencing batch reactor; SRED: slow release electron donor; CBP: carbohydrate base polymers; L: lignocellulose

In **Chapter 3**, the robustness and resilience of sulphate reduction to fest famine (FF) conditions was studied in inverse fluidized bed bioreactors (IFBB). It was found that the hydrodynamic regime directly influences the robustness, resilience and adaptation time of the IFBB, also, the COD:sulphate ratio was identified as the major factor affecting the BSR process. Furthermore, the effect of high rate (HRT $\leq$ 0.25 d) feeding conditions was studied in **Chapter 4** using an IFBB for biological sulphate reduction; the sulphate removal efficiency (SRE) was not affected by the low HRT $\leq$ 0.25 d tested, however, the COD loading rate did affect the biological performance.

In **Chapter 5**, the effect of the initial sulphate concentration was studied on the start up of BSR using a constant COD:sulphate ratio (2.4) in sequencing batch bioreactors (SBR); wherein the adaptation time and the BSR influenced the accumulation of either propionate or acetate on the start-up phase in an SBR. Whereas the influence of the initial COD, NH$_4^+$ and sulphate concentrations on sulphate reduction was studied in **Chapter 6** using batch bioreactors (BB), little NH$_4^+$ effect was observed on the sulphate removal rates and a major influence was observed on the electron donor uptake during sulphate reduction, also, the electron donor

5

utilization via the sulphate reduction process improved simultaneously to the decreasing initial electron donor concentrations.

In **Chapter 7**, the robustness of BSR was tested in SBR at a constant COD:sulphate ratio (2.4) but different COD and sulphate concentrations, apart from transient feeding conditions, the effect of NH_4^+ famine conditions was included in this study. The **Chapter 8** reports the research on the use of carbohydrate based polymers (CBP) as slow release electron donors (SRED) for BSR in batch bioreactors; the BSR process carried out at different rates using fermented products of the hydrolysis fermentation, the highest removal efficiencies were observed using cellulose as CBP. Afterwards, lignocellulosic polymers (L) were studied as SRED in batch and IFBB in **Chapter 9**; the findings in this chapter suggest the SRED hydrolysis-fermentation as the limiting step for BSR.

Finally, **Chapter 10** highlights and discusses the key points of the research performed along the PhD, and also discusses about future possible research work related to BSR process.

1.3   References

Bai, H., Kang, Y., Quan, H., Han, Y., Sun, J., Feng, Y., 2013. Treatment of acid mine drainage by sulfate reducing bacteria with iron in bench scale runs. Bioresour. Technol. 128, 818–822. doi:10.1016/j.biortech.2012.10.070

Borrego, J., Carro, B., López-González, N., de la Rosa, J., Grande, J.A., Gómez, T., de la Torre, M.L., 2012. Effect of acid mine drainage on dissolved rare earth elements geochemistry along a fluvial–estuarine system: the Tinto-Odiel Estuary (S.W. Spain). Hydrol. Res. 43, 262–274. doi:10.2166/nh.2012.012

Deng, D., Lin, L.-S., 2013. Two-stage combined treatment of acid mine drainage and municipal wastewater. Water Sci. Technol. 67, 1000–1007. doi:10.2166/wst.2013.653

Equeenuddin, S.M., Tripathy, S., Sahoo, P.K., Panigrahi, M.K., 2010. Hydrogeochemical characteristics of acid mine drainage and water pollution at Makum Coalfield, India. J. Geochemical Explor. 105, 75–82. doi:10.1016/j.gexplo.2010.04.006

Foucher, S., Battaglia-Brunet, F., Ignatiadis, I., Morin, D., 2001. Treatment by sulfate-reducing bacteria of Chessy acid-mine drainage and metals recovery. Chem. Eng. Sci. 56, 1639–1645. doi:10.1016/S0009-2509(00)00392-4

Hulshoff Pol, L.W., Lens, P.N.L., Stams, A.J.M., Lettinga, G., 1998. Anaerobic treatment of sulphate-rich wastewaters. Biodegradation 9, 213–224. doi:10.1023/A:1008307929134

Jiménez-Rodríguez, A.M., Durán-Barrantes, M.M., Borja, R., Sánchez, E., Colmenarejo, M.F., Raposo, F., 2009. Heavy metals removal from acid mine drainage water using biogenic hydrogen sulphide and effluent from anaerobic treatment: Effect of pH. J. Hazard. Mater. 165, 759–765. doi:10.1016/j.jhazmat.2008.10.053

Kage, S., Ikeda, H., Ikeda, N., Tsujita, A., Kudo, K., 2004. Fatal hydrogen sulfide poisoning at a dye works. Leg. Med. 6, 182–186. doi:10.1016/j.legalmed.2004.04.004

Lin, C., Wu, Y., Lu, W., Chen, A., Liu, Y., 2007. Water chemistry and ecotoxicity of an acid mine drainage-affected stream in subtropical China during a major flood event. J. Hazard. Mater. 142, 199–207. doi:10.1016/j.jhazmat.2006.08.006

Mapanda, F., Nyamadzawo, G., Nyamangara, J., Wuta, M., 2007. Effects of discharging acid-mine drainage into evaporation ponds lined with clay on chemical quality of the surrounding soil and water. Phys. Chem. Earth, Parts A/B/C 32, 1366–1375. doi:10.1016/j.pce.2007.07.041

Martins, M., Faleiro, M.L., Barros, R.J., Veríssimo, A.R., Barreiros, M.A., Costa, M.C., 2009. Characterization and activity studies of highly heavy metal resistant sulphate-reducing bacteria to be used in acid mine drainage decontamination. J. Hazard. Mater. 166, 706–713. doi:10.1016/j.jhazmat.2008.11.088

Monterroso, C., Macías, F., 1998. Drainage waters affected by pyrite oxidation in a coal mine in Galicia (NW Spain): Composition and mineral stability. Sci. Total Environ. 216, 121–132. doi:10.1016/S0048-9697(98)00149-1

Mulopo, J., Radebe, V., 2012. Recovery of calcium carbonate from waste gypsum and utilization for remediation of acid mine drainage from coal mines. Water Sci. Technol. 66, 1296–1300. doi:10.2166/wst.2012.322

Muyzer, G., Stams, A.J.M., 2008. The ecology and biotechnology of sulphate-reducing bacteria. Nat. Rev. Microbiol. 6, 441–454. doi:10.1038/nrmicro1892

Nieto, J.M., Sarmiento, A.M., Olías, M., Canovas, C.R., Riba, I., Kalman, J., Delvalls, T.A., 2007. Acid mine drainage pollution in the Tinto and Odiel rivers (Iberian Pyrite Belt, SW Spain) and bioavailability of the transported metals to the Huelva Estuary. Environ. Int. 33, 445–455. doi:10.1016/j.envint.2006.11.010

Reese, B.K., Anderson, M.A., Amrhein, C., 2008. Hydrogen sulfide production and volatilization in a polymictic eutrophic saline lake, Salton Sea, California. Sci. Total Environ. 406, 205–218. doi:10.1016/j.scitotenv.2008.07.021

Sarmiento, A.M., Olías, M., Nieto, J.M., Cánovas, C.R., Delgado, J., 2009. Natural attenuation processes in two water reservoirs receiving acid mine drainage. Sci. Total Environ. 407, 2051–2062. doi:10.1016/j.scitotenv.2008.11.011

Strosnider, W.H., Nairn, R.W., 2010. Effective passive treatment of high-strength acid mine drainage and raw municipal wastewater in Potosí, Bolivia using simple mutual incubations and limestone. J. Geochemical Explor. 105, 34–42. doi:10.1016/j.gexplo.2010.02.007

Vollertsen, J., Nielsen, A.H., Jensen, H.S., Wium-Andersen, T., Hvitved-Jacobsen, T., 2008. Corrosion of concrete sewers—The kinetics of hydrogen sulfide oxidation. Sci. Total Environ. 394, 162–170. doi:10.1016/j.scitotenv.2008.01.028

Yi, Z.-J., Tan, K.-X., Tan, A.-L., Yu, Z.-X., Wang, S.-Q., 2007. Influence of environmental factors on reductive bioprecipitation of uranium by sulfate reducing bacteria. Int. Biodeterior. Biodegradation 60, 258–266. doi:10.1016/j.ibiod.2007.04.001

Chapter 2

Literature review

A modified version of this chapter was published as:
Reyes-Alvarado L.C., Rene E.R., Esposito G., and Lens P.N.L., (2018). Bioprocesses for sulphate removal from wastewater. In: Varjani S., Gnansounou E., Gurunathan B., Pant D., Zakaria Z. (eds.) Waste Bioremediation. Energy, Environment and Sustainability. Springer Nature Singapore Pte Ltd. doi: 10.1007/978-981-10-7413-4_3

Abstract

This chapter highlights and discusses the important research studies that have been carried out previously on the sulphate removal from wastewaters under anaerobic conditions. Moreover, the role of electron donor addition on biological sulphate reduction and the beneficial role of sulphate reducing bacteria (SRB) are reviewed in this chapter. Briefly stating, this chapter describes the fundamentals of anaerobic digestion, the sulphate reduction process, the factors affecting biological sulphate reduction and the different bioreactor configurations used for sulphate reduction in wastewater. Besides, kinetic modeling and artificial neural network based modeling literature is also reviewed.

Keywords: biological sulphate reduction; sulphidogenesis; bioreactors; electron donors; artificial neural networks; sulphate reducing bacteria; sulphate rich wastewater

2.1 Anaerobic digestion

The anaerobic digestion (AD) process involves the decomposition of organic matter by a consortium of microbes in an environment free of oxygen (Molino *et al.*, 2013; Ward *et al.*, 2008). It involves the stabilization and degradation of organics under anaerobic conditions by microorganisms, leading to the formation of microbial biomass and biogas (a mixture of carbon dioxide and methane) (Chen *et al.*, 2008; Kelleher *et al.*, 2002). Table 2-1 shows the different stoichiometric equations governing the anaerobic degradation process and the corresponding Gibbs free energy values ($\Delta G^{0\prime}$). Among the different biological treatment methods, AD is frequently used due to its cost effectiveness, high energy recovery potential and limited environmental impacts (Mata-Alvarez *et al.*, 2000). The AD process is made up of a hydrolysis-fermentation, acetogenesis and the methanogenesis phase.

The successful implementation of this technology has been envisaged in the treatment of food wastes, wastewater sludge and agricultural wastes. The reduction of biological oxygen demand (BOD), chemical oxygen demand (COD) and the production of renewable energy in waste streams are some of the advantages of this technology (Chen *et al.*, 2008). The success of AD processes is governed by carefully controlling the operating parameters, especially, the reactor configurations (batch or continuous), operating temperatures (psychrophilic, mesophilic or thermophilic), reactor design (plug-flow or completely mixed), and solid content (wet or dry) (Li *et al.*, 2011).

2.1.1 Hydrolysis-fermentation

Hydrolytic-fermentative bacteria (HFB) hydrolyze and ferment organic matter. Organic polymeric materials are hydrolyzed to monomers, *e.g.* cellulose to sugars, oil and fat to fatty acids and proteins to amino acids, by hydrolytic enzymes (cellulases, amylases, proteases and lipases) which are secreted by microorganisms (Molino *et al.*, 2013). In the second stage, named acidogenic phase or fermentation, the monomers are then converted by HFB to a mixture of short-chain volatile fatty acids (VFAs) such as butyric acid ($CH_3CH_2CH_2COOH$), propionic acid (CH_3CH_2COOH) and acetic acid (CH_3COOH) mainly, but also ethanol (CH_3CH_2OH) can be produced, as *e.g.* Eq. 2-1 to Eq. 2-6 (Table 2-1).

2.1.2 Acetogenesis

Acetogenic bacteria, also known as acid formers, convert simple organic acids to acetate, CO_2 and H_2 during the acetogenesis step (*e.g.* Eq. 2-7 to Eq. 2-13). Microorganisms such as Syntrophobacter wolinii (a propionate decomposer) are responsible for the products formed during acetogenesis. Different acid formers such as *Sytrophomonos wolfei* (a butyrate decomposer) and species of *Clostridium, Peptococcusanerobus, Lactobacillus* and *Actinomyces* are also involved in the acetogenesis step (Li *et al.*, 2011).

Table 2-1. Stoichiometric reactions in an anaerobic degradation process

Hydrolysis:	$\Delta G^{0\prime}$ (kJ.reaction^{-1})	
$(C_6H_{10}O_5)_n + {}_nH_2O \rightarrow {}_nC_6H_{12}O_6$		Eq. 2-1
Fermentation:		
$C_6H_{12}O_6 + 2\ H_2O \rightarrow CH_3(CH_2)_2COO^- + 2\ HCO_3^- + 3\ H^+ + 2\ H_2$	-254.8	Eq. 2-2
$C_6H_{12}O_6 + 2\ H_2 \rightarrow 2\ CH_3CH_2COO^- + 2\ H_2O + 2\ H^+$	-358.1	Eq. 2-3
$C_6H_{12}O_6 \rightarrow CH_3CHOHCOO^- + 2\ H^+$	-198.3	Eq. 2-4
Glycerol $\rightarrow$ Pyruvate$^-$ + H$^+$ + 2H$_2$	-25.9	Eq. 2-5
$3\ CH_3CHOHCOO^- \rightarrow 2\ CH_3CH_2COO^- + CH_3COO^- + CO_2$	-54.9	Eq. 2-6
Acetogenesis		
$C_6H_{12}O_6 + 4\ H_2O \rightarrow 2\ CH_3COO^- + 2\ HCO_3^- + 4\ H^+ + 4\ H_2$	-206.3	Eq. 2-7
Pyruvate$^-$ + 2 H$_2$O $\rightarrow$ CH$_3$COO$^-$ + HCO$_3^-$ + H$^+$ + H$_2$	-47.3	Eq. 2-8
$CH_3(CH_2)_2COO^- + 2\ H_2O \rightarrow 2\ CH_3COO^- + H^+ + 2\ H_2$	+48.1	Eq. 2-9
$CH_3CH_2COO^- + 3\ H_2O \rightarrow CH_3COO^- + HCO_3^- + H^+ + 3\ H_2$	+76.1	Eq. 2-10
$CH_3CHOHCOO^- + 2\ H_2O \rightarrow CH_3COO^- + HCO_3^- + H^+ + 2\ H_2$	-4.2	Eq. 2-11
$CH_3CH_2OH + H_2O \rightarrow CH_3COO^- + H^+ + 2\ H_2$	+9.6	Eq. 2-12
$4\ H_2 + 2\ HCO_3^- + H^+ \rightarrow 4\ H_2O + CH_3COO^-$	-104.5	Eq. 2-13
Methanogenesis		
$CH_3COO^- + H_2O \rightarrow CH_4 + HCO_3^-$	-31.1	Eq. 2-14
$4\ H_2 + HCO_3^- + H^+ \rightarrow CH_4 + 3\ H_2O$	-135.6	Eq. 2-15

2.1.3 Methanogenesis

During methanogenesis, methane is produced by methanogenic Archaea in two steps: either through the division of acetate molecules to generate HCO$_3^-$ and CH$_4$ (Eq. 2-14), or by the reduction of HCO$_3^-$ in the presence of H$_2$ (Eq. 2-15). The production of methane is higher from the reaction involving the reduction of carbon dioxide (hydrogenotrophic methanogens) when compared to the reaction from the cleavage of acetate (acetoclastic methanogens), although digesters with hydrogen limiting conditions generates more acetate (Molino *et al.*, 2013).

Once produced, the biogas is generally composed of ca. 48 to 65% methane, ca. 36 to 41% carbon dioxide, up to 17% nitrogen, < 1% oxygen, 32 to 169 ppm hydrogen sulphide, and traces of other gases (Rasi *et al.*, 2007).

2.2   The sulphate reduction process

2.2.1   Sulphur cycle

The tenth most abundant element on the surface of the earth is sulphur. Organisms require it for processing of vitamins, amino acids and hormones. Microbes play a major role in the biogeochemical sulphur cycle. The sulphur oxidation states are -2 (sulphide), 0 (elemental sulphur) and +6 (sulphate), among which sulphate is very important for nature. Sulphur biogeochemical pathways are known to interact with those of other elements, especially metals (Pepper *et al.*, 2004).

Wastewater with high sulphate concentrations are produced through leaching from landfills (Nedwell and Reynolds, 1996), this can cause an unbalance to the natural biological sulphur cycle by altering biodegradation pathways and the kinetic rates. A different oxidation state (-2 and +6) of sulphur can be found in other sources such as wastewater from the textile industry. Cirik *et al.* (2013) reported that sulphate in textile industries is added to dye baths for ionic strength regulation. Deep sea venting, volcanic activity, bacterial activities, fossil combustion, and industrial emissions are some of the major sources of sulphate in the atmosphere. Sulphate oxidized from sulphur in the atmosphere can be deposited in wet or dry form.

Redox reactions generally characterizes the sulphur cycle (Figure 2-1), sulphur can be reduced to sulphide, which in turn can be oxidized to elemental sulphur or sulphate by microbes. However, sulphate can be reduced back to sulphide by sulphate reducing bacteria (Robertson and Kuenen, 2006). The release of sulphur aerobically or anaerobically from its organic form is known as sulphur mineralization.

Chemoautotrophic bacteria, heterotrophic microorganisms and fungi under aerobic conditions oxidize sulphur to sulphate or thiosulphate. Phototropic or chemolitithothrophic bacteria fix CO_2 by utilizing light energy in oxidizing sulphide to sulphur or sulphate. When there is an imbalance in the reductive or oxidative paths, an accumulation of intermediates such as elemental sulphur, iron sulphide and hydrogen sulphide occurs. The process of sulphur disproportionation is an energy generating process carried out by sulphur reducing bacteria

(SRB), wherein elemental sulphur or thiosulphate acts as both electron donor and acceptor, and results in the formation of sulphate and sulphide, respectively (Tang *et al.*, 2009).

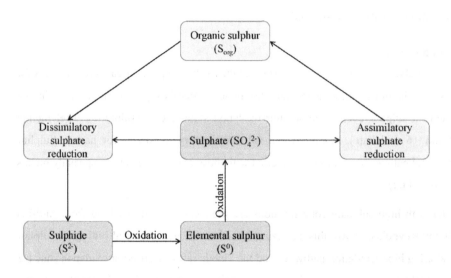

Figure 2-1. The biological sulphur cycle

2.2.2 Biological sulphate reduction

Two anaerobic microbial degradation pathways (Figure 2-2) are well documented in the literature. The sulphate removal can be assimilatory or dissimilatory (Figure 2-3). In the assimilatory pathway, sulphate is reduced to sulphide, in small quantities, and later the sulphide is converted to cysteine. This amino acid is the source of other biological sulphur containing molecules (Leustek *et al.*, 2000). The dissimilatory pathway is confined to two archaeal and five bacterial genera, wherein the terminal electron acceptor (sulphate) produces large quantities of sulphide and the process is also known as sulphidogenesis (Grein *et al.*, 2013). The two pathways have a similar starting point: the activation of sulphate by reaction with adenosine-5'-Triphosphate (ATP) forming adenosine-5'-phosphosulphate (APS). Sulphate adenylyl transferase (SAT) acts as a catalyst in this step, also referred to as ATP sulphurylase (Taguchi *et al.*, 2004).

Sulphate is reduced using an electron donor to produce sulphide and sulphate reducing bacteria (SRB) responsible of this this process (Hao *et al.*, 1996). Sulphate reduction (Eq. 2-16 to Eq. 2-29) using electron donors such as lactate, propionate, acetate and hydrogen is summarized in Table 2-2.

The initial step of biological sulphate reduction involves the transfer of exogenous sulphate through the bacterial cell membrane into the cell. The sulphidogenesis step proceeds via the action of ATP sulphurylase after arriving into the cell membrane (Figure 2-3). ATP produces the highly activated molecule APS, and pyrophosphate (PPi) in the presence of sulphate, which may be attached to yield inorganic phosphate. APS is rapidly converted to sulphite (SO_3^-) by the cytoplasmic enzyme APS reductase. Sulphite, in turn may be reduced via a number of intermediates to form the sulphide ion. The physiology and growth of SRB has been studied and well documented in the literature (Hao *et al.*, 1996; Matias *et al.*, 2005; Muyzer and Stams, 2008; Rabus *et al.*, 2006; Zhou *et al.*, 2011).

Table 2-2. Stoichiometric reactions involved in sulphidogenesis

Sulphidogenesis	$\Delta G^{0\prime}$ (kJ.reaction^{-1})	
Glucose + SO_4^{2-} → 2 CH_3COO^- + HS^- + 2 HCO_3^- + 3 H^+	-358.2	Eq. 2-16
2 $CH_3CHOHCOO^-$ + SO_4^{2-} → 2 CH_3COO^- + HS^- + 2 HCO_3^- + H^+	-160.1	Eq. 2-17
2 $CH_3CHOHCOO^-$ + 3 SO_4^{2-} → 6 HCO_3^- + HS^- + H^+	-255.3	Eq. 2-18
$CH_3(CH_2)_2COO^-$ + 0.5 SO_4^{2-} → 2 CH_3COO^- + 0.5 HS^- + 0.5 H^+	-27.8	Eq. 2-19
$CH_3(CH_2)_2COO^-$ + 3 SO_4^{2-} + 2 H_2 → CH_3COO^- + HS^- + HCO_3^- + 2 H_2O	-198.4	Eq. 2-20
$CH_3CH_2COO^-$ + 0.75 SO_4^{2-} → CH_3COO^- + HCO_3^- + 0.75 HS^- + 0.25 H^+	-37.7	Eq. 2-21
$CH_3CH_2COO^-$ + 1.75 SO_4^{2-} → 3 HCO_3^- + 1.75 HS^- + 0.25 H^+	-85.4	Eq. 2-22
$CH_3CH_2COO^-$ + SO_4^{2-} + H_2 → CH_3COO^- + HS^- + HCO_3^- + 2 H_2O	-75.8	Eq. 2-23
2 CH_3CH_2OH + SO_4^{2-} → 2 CH_3COO^- + HS^- + H^+ +H_2O	-22	Eq. 2-24
2 CH_3OH + SO_4^{2-} → 2 $HCOO^-$ + HS^- + H^+ + 2 H_2O	-108.3	Eq. 2-25
CH_3COO^- + SO_4^{2-} → 2 HCO_3^- + HS^-	-48	Eq. 2-26
$HCOO^-$ + SO_4^{2-} + H^+ → HS^- + 4 HCO_3^-	-144	Eq. 2-27
4 CO + SO_4^{2-} + 4 H_2O → HS^- + 4 HCO_3^- + 3 H^+	-212	Eq. 2-28
4 H_2 + SO_4^{2-} + H^+ → HS^- + 4 H_2O	-151.9	Eq. 2-29

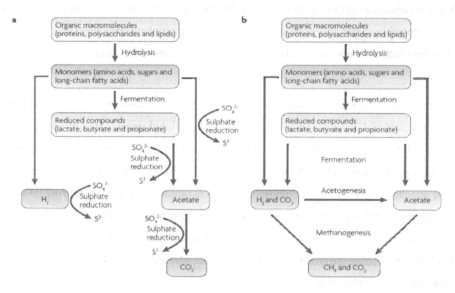

Figure 2-2. Pathway for anaerobic degradation of organic substrates: a) sulphidogenesis, and b) methanogenesis (Adapted from Muyzer and Stams, 2008)

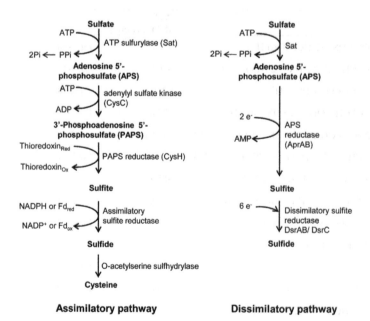

Figure 2-3. The prokaryotic assimilatory and dissimilatory pathways for sulphate reduction (Adapted from Grein et al., 2013)

2.2.3   Sulphate reducing bacteria (SRB)

SRB can be categorized into two classes depending on their biodegradation potential: those leading to incompletely degradable organic compounds forming acetate and those completely degrading organics to CO_2 (Muyzer and Stams, 2008). The availability of substrate is sometimes affected by the competition between SRB and the methonogens. Several factors, among others, facilitate SRB in outcompeting methanogens. These factors include anaerobic respiration in the presence of sulphate as the final electron acceptor leading to more energy for the growth of SRB and at conditions difficult for methanogens. In addition, SRB are able to consume substrates to very low concentrations because of their high affinity for hydrogen and acetate (Rabus *et al.*, 2006). It is noteworthy to mention that SRB have a higher specific growth rate compared to methanogens (Moestedt *et al.*, 2013). *Desulfovibrio* species have high affinity for hydrogen and is contemplated to be the rational for outcompeting hydrogenotrophic methanogens under sulphidogenic conditions (Widdel, 2006). *Desulfobacter, Desulfobulbus, Desulfococcus, Desulfocarcina, Desulfomaculum, Desulfonema*, and *Desulfovibro* use sulphate as the terminal electron acceptor, using acetate, lactate and methanol as the electron donor (Pepper *et al.*, 2004). Polymeric compounds such as protein, starch and cellulose are not utilized directly by SRB as the substrates, but they depend on other microorganisms to ferment these compounds to products which can be used as substrates by the SRB. An analysis of 16S ribosomal ribonucleic acid (rRNA) by Muyzer and Stams (2008) grouped SRB into seven different lineages, two of which were archaea and five were bacteria (Figure 2-4).

The ecology, bioenergetics, and physiology of SRB have been discussed in a number of review articles (Barton and Fauque, 2009; Gibson, 1990). SRB are known to exist in different environments such as: anoxic estuarine sediment, acid mine water, saline water, freshwater and generally in all soil types. The temperature range at which they grow is also diverse and thermophilic SRB have been isolated at temperatures > 60°C in deep aquifers (Hao *et al.*, 1996). According to Mizuno *et al.* (1998) hydrogen-consuming and lactate-consuming SRB can be enumerated using the most probable (MPN) counts technique, while qualitatively, the presence of SRB can be confirmed by the presence of black FeS precipitates.

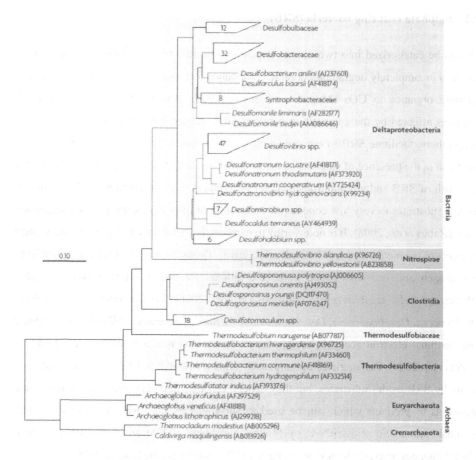

Figure 2-4. Phylogenetic tree on 16S ribosomal RNA(rRNA) sequence of SRB species (Adapted from Muyzer and Stams, 2008)

2.3   Electron donors for SRB

Al-Zuhair *et al*. (2008) determined appropriate sulphate concentration for SRB when grown as a pure culture. Several electron donors have been studied as energy and carbon sources for SRB (Table 2-3). Most electron donors are products from fermentation, monomers or cell components from other sources. Hydrogen is one of the most important substrates for SRB. *Desulfovibrio* species has a high affinity for hydrogen, and is considered to be the reason why they are able to out-compete hydrogenotrophic methanogens in sulphate rich environments (Widdel, 2006).

Table 2-3. Compounds used as energy substrates by SRB (Adapted from Hansen, 1993)

Compound	Substrates
Inorganic	Hydrogen, carbon monoxide
Monocarboxylic acids	Lactate, acetate, butyrate, formate, propionate, isobutyrate, 2- and 3-methylbutyrate, higher fatty acids up to C18, pyruvate
Dicarboxylic acids	Succinate, fumarate, malate, oxalate, maleinate, glutarate, pimelate
Alcohols	Methanol, ethanol, propanol, butanol, ethylene glycol, 1,2- and 1,3-propanediol, glycerol
Amino acids	Glycine, serine, cysteine, threonine, valine, leucine, isoleucine, aspartate, glutamate, phenylalanine
Miscellaneous	Choline, furfural, oxamate, fructose, benzoate, 2-, 3- and 4-OH-benzoate, cyclohexanecarboxylate, hippurate, nicotinic acid, indole, anthranilate, quinoline, phenol, p-cresol, catechol, resorcinol, hydroquinone, protocatechuate, phloroglucinol, pyrogallol, 4-OH-phenyl-acetate, 3-phenylpropionate, 2-aminobenzoate, dihydroxyacetone

Furthermore, SRB require nitrogen, phosphorus and iron. Recent energy sources for biological sulphate reduction include complex organic carbon sources. Sewage sludge was one of the first carbon sources considered because of its complexity (Butlin et al., 1956). Van Houten et al. (1996) studied the use of synthesis gas (mixtures of $H_2/CO/CO_2$) as energy source and the stimulation of biological sulphate reduction in the presence of SRB utilizing complex organic carbon sources. Numerous organic waste matrices have also been used as carbon sources and electron donors. These include mushroom, leaf mulch, wood chips, sewage sludge, sawdust, compost, animal manure, whey, vegetable compost, and other agricultural waste (Liamleam and Annachhatre, 2007).

2.3.1 Organic solids

2.3.1.1 Starch

Potato is a staple food in Europe and other parts of the world. It is also produced in large quantities in The Netherlands and according to a recent report, the estimated potato production in the year 2011 was ~ 73 million metric tons (FAOSTAT, 2011). Table 2-4 shows the composition of fresh potato. They contain ~ 70-80% water and starch counts for 16-24% of the total weight. Potato is considered to be the fourth most important food crop in the world after wheat, rice and maize. Kang and Weiland (1993) ascertained the biodegrading characteristics of potato using batch tests and reported that ~ 90% potato could be degraded at 35°C at a substrate to inoculum ratio of 0.8.

Table 2-4. Proximate composition of different potatoes (Adapted from Hoover, 2001)

Starch source	Starch yield (%)	Size (μm) and shape	Amylose content (%)	Total lipid (%)	Phosphorus (% dsb)		Nitrogen (%)
					Org	Inorg	
Solanum tuberosum (potato)	32	15-110 oval, spherical	25.4	0.19	0.089	0.001	0.1
Ipomea babatas (sweet potato)	30	2-42 round, oval and polygonal	19.1	0.06-0.6	0.012	-	0.006
Solanum tuberosum (waxy potato)	-	14-44 round, oval	-	-	0.069	0.001	-

Note: dsb-double strand breaks

2.3.1.2 Cellulose

Cellulose exists in abundance on earth under different forms. The anaerobic degradation of cellulose begins with depolymerization and is followed by solubilization. Besides, degradation products of cellulose (*i.e.* celloboise) can be converted to CH_4 and CO_2 through acidogenesis, acetogenesis, and methanogenesis, respectively. When anaerobic microorganisms excrete cellulosomes from their cell wall, which are then attached to the cellulose particles, this is termed as hydrolysis. Several studies have also argued whether the bacterial hydrolysis or methanogenesis phase is the rate-limiting step for biotransformation of polymeric compounds rich in cellulose content (Jeihanipour *et al.*, 2011). Recent studies have also demonstrated the anaerobic digestion of cellulose under mesophilic and thermophilic conditions (O'Sullivan *et al.*, 2008; Xia *et al.*, 2012). According to Yang *et al.* (2004), in batch experiments, thermophilic cellulose digestion is less effective compared to the mesophilic conditions. In that study, 16% volatile solids were removed under thermophilic conditions, while 52% volatile solids were removed under mesophilic conditions in 30 d.

2.3.1.3 Proteins

The hydrolysis products of proteins under anaerobic conditions include peptides and amino acids, which are further degraded to ammonium, carbon dioxide, short-chain fatty acids, and hydrogen. According to Örlygsson *et al.* (1994), the hydrolysis of protein is frequently affected by the electron donor availability. The hydrolysis rate under anaerobic conditions is lower than that observed under aerobic conditions. Deamination is the initial step of the degradation of protein, which is favoured under aerobic rather than anaerobic conditions (Shao *et al.*, 2013). Only a few members of the *Desulfobacterium*, *Desulfotomaculum* and *Desulfovibrio* genera utilize amino acids. Baena *et al.* (1998) reported that the addition of thiosulphate to growth media enabled to optimize the degradation of amino acids by an SRB (Desulfovibrio aminophilus sp. nov. DSM 12254), indicating the vital role played by sulphate and thiosulphate in the degradation of proteinaceous compounds.

2.3.1.4 Chitin

Chitin is a polymer, occurring naturally as a white, hard, inelastic and nitrogenous polysacharide. It can be found in the exoskeletons of crabs and shrimps in the alpha-chitin form (Rinaudo, 2006). The beta-chitin form has a higher affinity for solvents than the alpha form because it has weak hydrogen bonds (Pillai *et al.*, 2009). One of the derivatives of chitin is chitosan and is obtained by alkaline deacetylation of chitin in a strong alkaline solution.

2.3.2 Selection of electron donors for biological sulphate reduction

According to van Houten *et al.* (1996), the three major criteria for selecting a suitable electron donor for the sulphate reduction process are: (i) high sulphate removal efficiency complemented by a low COD effluent concentration; (ii) electron donor availability, and (iii) reduced cost of sulphate unit converted to sulphide. Thermodynamic parameters such as physiological free energies $\Delta G^{0'}$ (kJ.mol^{-1}) of the sulphidogenesis (Table 2-2) are also important for treatment efficiency.

Table 2-5. Sulphate reduction at different operational conditions and different bioreactors (Adapted from Liamleam and Annachhatre, 2007)

Electron donors	Temperature (°C)	Reactor type	SO_4^{2-} removal rate (g. $L^{-1}d^{-1}$)
Acetate	35	Packed bed	15-20
Acetate	33-35	EGSB	28.5
CO	35	Pack bed	2.4
Ethanol	33	EGSB	21
Glucose/ Acetate	35	Anaerobic digester	1.92
H_2/CO_2	30	Gas-lift	30
H_2/CO	35	Pack bed	1.2
Lactate	Room temp	Plug flow	0.41
Molasses	30	UASB	4.3
Molasses	27	CSTR	0.84
Molasses	35	Anaerobic RBC	0.35
Molasses	31	Packed bed	6.5
Synthesis gas	30	Gas lift	12-14

Note: CO-carbon monoxide; EGSB-Expanded granular sludge bed; UASB-Upflow anaerobic sludge blanket; RBC-Rotating biological contactor; CSTR-Continuous stirred tank reactor

2.3.2.1 Efficiency of sulphate removal

The information presented in Table 2-5 indicates that acetate (28.5 g. $L^{-1}d^{-1}$), ethanol (21 g. $L^{-1}d^{-1}$), and hydrogen (51 mmol. $L^{-1}d^{-1}$) have the highest sulphate removal rates. However, lactate (0.41 g. $L^{-1}d^{-1}$) and molasses (< 6.5 g. $L^{-1}d^{-1}$) have lower sulphate removal rates. The electron donors summarized in Table 2-5 indicate different affinities for the carbon source by SRB (Stams *et al.*, 2005). Additionally, the bioreactor configuration plays an important role in determining the sulphate reduction efficiency (Kijjanapanich *et al.*, 2014). The key factor affecting the sulphidogenesis is the capability of retaining the active SRB in the bioreactor.

2.3.2.2 Availability and cost of electron donor

Lactate and molasses, though cost-effective, are not completely oxidized by SRB, generating high COD concentrations in the effluent (Liamleam and Annachhatre, 2007). Hydrogen and ethanol are not cost-effective, but are still used for sulphate loads exceeding 200 kg $SO_4^{2-}.h^{-1}$. However, due to safety reasons, ethanol is preferred over hydrogen.

2.3.3 Environmental parameters affecting sulphate reduction

2.3.3.1 Temperature

A very important factor in the AD process is temperature, and the performance of bioreactors usually varies depending on the operating temperatures and the adaptability of the microbes to different temperature ranges Table 2-6. SRB comprise both mesophilic and thermophilic strain which are affected by temperature. Weijma *et al.* (2000) showed a significant increase in sulphate reduction when temperature was increased from 20 to 35°C, but bacterial activity decreased at 40°C. According to Tsukamoto *et al.* (2004), the efficiency of acid mine drainage treatment was not affected by temperature due to acclimatization of SRB to low temperature conditions over prolonged time. Thermophilic processes lead to H₂S stripping, thereby reducing the concentration in the liquid phase. Therefore, the treatment of sulphate rich wastewater is made more efficient at temperatures of 55-70°C (Sarti and Zaiat, 2011).

Table 2-6. Temperature range for the growth of a number of SRB (Adapted from Tang *et al.*, 2009)

SRB	Temperature (°C)	
	Range	Optimum
Desulfobacter	28-32	
Desulfobulbus	28-39	
Desulfomonas	-	30
Desulfosarcina	33-38	
Desulfovibrio	25-35	
Thermodesulforhabdus norvegicus	44-74	60
Desulfotomaculum luciae	50-70	
Desulfotomaculum solfataricum	48-65	60
Desulfotomaculum thermobenzoicum	45-62	55
Desulfotomaculum thermocisternum	41-75	62
Desulfotomaculum thermosapovorans	35-60	50
Desulfacinum infernum	64	-

2.3.3.2 pH and S²⁻ concentration

SRB show high specific activities in the pH range of 5.0 and 8.0. Beyond this range, their metabolic activity reduces and inhibition effects set in (Dvorak *et al.*, 1992). Sheoran *et al.* (2010) reported that some SRB are able to remove 38.3% of sulphate from the influent with a pH of 3.3, but their removal performance dropped when the pH was reduced to ~ 3.0. The hydrogen sulphide and bicarbonate present in the system buffers the solution pH, the buffering capacity depends on the type of organic end products, their composition and quantity (Dvorak et al., 1992). This causes inhibition of AD processes and could lead to failure of it. The effect of sulphide is believed to be caused by non-ionized H_2S, because neutral molecules can be permeated by cell membranes (Sarti and Zaiat, 2011).

2.3.3.3 Hydraulic retention time (HRT)

The HRT determines the time allowable for the SRB to adapt to the environment, initiate growth and metabolic activity thereby increasing the amount of sulphate or COD reduced. In bioreactors for sulphate reduction, a long HRT may lead to high sulphate reduction efficiencies and a complete oxidation of the electron donor used with a minimal residual acetate (Kaksonen et al., 2006). However, according to Sheoran et al. (2010), a short HRT may reduce the time available for the SRB to metabolize the substrate and could lead to complete washout of biomass from the reactor.

2.4   Conventional bioreactors for sulphate reduction

Several branches of biotechnology use bioreactors, such as biofuel production (Ozmihci and Kargi, 2008), food industries (Genari et al., 2003), production of pharmaceutical compounds (John et al., 2007) and environmental technologies (Show et al., 2011). Anaerobic wastewater treatment systems use mixed microbial consortia, which is somewhat different compared to other biotechnological process where isolation or/and sterilization is required (Goršek and Tramšek, 2008). Setting different steps of a process in one stage can make the process more attractive, in terms of process intensification. Therefore, the use of flocs, granules and biofilms is of great interest in biotechnology. This is possible by facilitating solid-liquid separation, and these coupled to the reactor configuration, make the separation of the three active phases (liquid-gas-solid) and downstream processing feasible. Granules and biofilms are easier to separate compared to other systems, and the use of settlers it is not necessary. Additionally, the surface area inside the reactor is increased; therefore, a large volume of diluted water can be treated.

Laboratory scale experiments have shown promising results for the treatment of wastewater rich in sulphate by using different bioreactor configurations. A variety of bioreactors (Figure 2-5) such as expanded granular sludge bed reactors (EGSB), fluidized bed reactors (FBR), gas-lift bioreactors (GLB), inverse fluidized bed reactors (IFB), membrane bioreactors (MBR), sequencing batch reactors (SBR), and upflow anaerobic sludge blanket (UASB) have been used for sulphate reduction in wastewaters (Sheoran *et al.*, 2010). Numerous two-stage processes combining anaerobic biological sulphate reduction with an aerobic step have also been used at laboratory-scale (du Preez *et al.*, 1992; Maree and Hill, 1989). Up flow packed bed reactors (Fontes Lima and Zaiat, 2012; Peixoto *et al.*, 2011), stirred tank reactors (Kieu *et al.*, 2011), sulphate reducing columns (Baskaran and Nemati, 2006) and biofilms (D′Acunto *et al.*, 2011) have also been studied. These bioreactors have shown efficient sulphate reduction efficiencies alongside selective removal of heavy metals from effluents by sulphide precipitation and pH manipulation (Sampaio *et al.*, 2010; Villa-Gomez *et al.*, 2011).

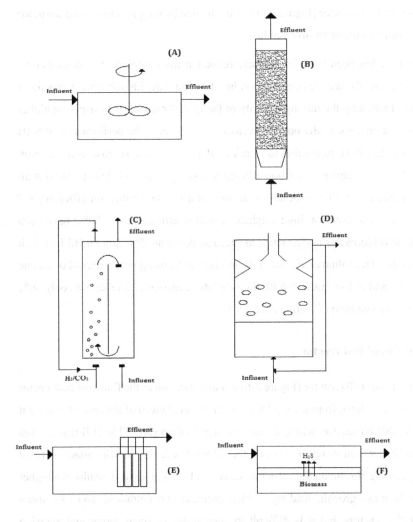

Figure 2-5. Schematic representation of bioreactors (Adapted from Papirio *et al.*, 2013)
The CSTR (A), PBR (B), GLB (C), UASB bioreactor (D), immersed membrane bioreactor (IMBR) (E) and an extractive membrane bioreactor (EMBR) (F)

2.4.1 UASB bioreactor

The UASB bioreactor is an invention of Gatze Lettinga. It is a mature technology for wastewater treatment (Lier *et al.*, 2015). The UASB bioreactor is operated at upward velocities < 2 m. h^{-1} (Hulshoff Pol *et al.*, 2004; van Haandel *et al.*, 2006). An UASB is considered as a high rate bioreactor because of its capability to deconvolute the solid retention time (SRT) from the HRT. The deconvolution of the HRT from the SRT is possible due to flocs and granules formation that ensure high biomass concentrations (Lier *et al.*, 2015; van Haandel *et al.*, 2006). UASB reactors are designed to handle a volumetric loading rate of 4 to 15 kg COD. m^3 d^{-1} (Lier *et al.*, 2015). The UASB bioreactor (Figure 2-5D) is intensified by the gas-liquid-solid separator placed at the top with a shape of an inverted funnel.

The UASB bioreactor has been used for sulphate reduction from sulphate rich wastewaters. Long HRT are beneficial for acetate consumption by SRB and it avoids substrate competition with methanogens. Increasing the mixing capacity of the UASB bioreactor by applying higher recirculation rates and increasing the upwards velocity can increase the performance of SRB (Arne Alphenaar *et al.*, 1993). According to Lopes *et al.* (2008), using sucrose as an electron donor and at pH < 7.0, the sulphate reduction efficiency was higher in an UASB (> 50%) at an HRT ~ 23 h compared to a CSTR bioreactor that showed a sulphate reduction efficiencies < 38% at an HRT ~ 20 h. In contrast, high sulphate reduction efficiencies (> 80%) have been reported for an UASB bioreactor using lactate as electron donor at 25°C and an HRT of 24 h after 500 d of operation (Bertolino *et al.*, 2012). In another study, using either ethanol or acetate as the carbon source and at low HRT (> 6 h), the sulphate reduction efficiency was only 30% at a limiting COD:sulphate ratio of 1 (Jing *et al.*, 2013).

2.4.2 Inverse fluidized bed reactor

An inverse fluidized bed (IFB) reactor (Figure 2-6) is a modification of the fluidized bed reactor where the liquid is recirculated from the top (downwards recirculation) of the reactor making it different from a UASB bioreactor, where liquid is recirculated upwards. The IFB reactor uses a carrier material lighter than water onto which the sulphate reducing biofilm attaches. In an IFB reactor, the growing biofilm on the carrier material is advantageous as it results in a higher surface area for biomass growth, leading to high biomass concentrations and low space requirements for the reactors, but it is difficult to control due to shear forces and abrasion (Davey and O'toole, 2000; Escudié *et al.*, 2011).

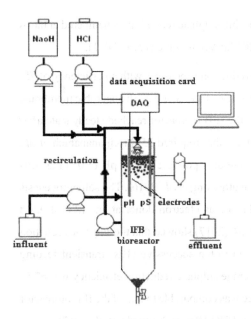

Figure 2-6. Schematic of an inverse fluidized bed reactor (Adapted from Villa-Gomez *et al.*, 2014)

Industrial and municipal wastewater treatment have been using biofilm based reactors since the last decades. Several studies have been done to establish the factors that affect biofilm formation and growth. In an IFB, the volumetric conversion rate of the pollutant depends on the liquid velocity and the substrate concentration. According to the results obtained by Eldyasti *et al.* (2012), the substrate concentration has a greater effect on diffusion rates than the liquid velocities, whereas Diez Blanco *et al.* (1995) showed contradictory results indicating a reduction in the external mass transfer velocity when the liquid velocity was increased. From a hydrodynamics point of view, Andalib *et al.* (2012) found that the diffusion rates were affected more by the liquid flow rate, *i.e.* under turbulent flow, rather than under laminar flow.

Table 2-7 summarizes the recent studies on sulfate reduction using IFB bioreactors. Recent studies on sulphate reduction use anaerobic sludge from methanogenic bioreactors as inoculum and low density polyethylene is the most preferred carrier material for IFB bioreactors (Table 2-7). Till date, sulphate reduction has been studied at the lowest HRT of 0.37 d. In a recent study by Villa-Gomez *et al.* (2011), the sulphate reduction efficiencies were 74 and 38% at a COD:sulphate ratio of 5 and 1, respectively. The influence of different electron donors on the sulphate reduction has been studied in the IFB bioreactor where the most commonly used electron donors are acetate, propionate, butyrate, ethanol and lactate. These electron donors have been studied at different COD:sulphate ratios, but sulphate reductions is hampered at ratios < 1.0 and optimal at ratios > 1.0 (Papirio *et al.*, 2013a; Villa-Gomez *et al.*, 2011).

Kijjanapanich *et al.* (2014) showed that sulphate reduction is possible with an efficiency of 75-85% regardless of the bioreactor configuration: *e.g.* IFB, UASB or gas lift anaerobic membrane bioreactor, and at a HRT of 0.64 d. In that study, the IFB bioreactor reached steady state after 20 d of operation compared to the UASB bioreactor that required 35 d (Kijjanapanich *et al.*, 2014). Sulphate reduction efficiencies of ~ 50% are also possible at low pH (5.0) in an IFB bioreactor and at a COD:sulphate ratio of 1 (Janyasuthiwong *et al.*, 2016). The sulphidogenesis is robust to transient feeding conditions using lactate as electron donor at an HRT of 0.5 d (Reyes-Alvarado *et al.*, 2017). Reyes-Alvarado *et al.* (2017) showed that the sulphate reduction is more affected by COD:sulphate ratio (< 1) than to ten successive (10x) transient feeding condition applied to the IFB. For instance, the average sulphate reduction efficiency was 67 ($\pm$ 15)% during the feast periods and this performance was comparable to that of the IFB bioreactor operation under normal operating conditions (61 $\pm$ 15%) (Reyes-Alvarado *et al.*, 2017).

2.4.3 Factors affecting bioreactor performance

2.4.3.1 Characteristics of organic substrate

Different organic substrates can be used as carbon source and electron donor for sulphate reduction (Table 2-3). However, the characteristics of the substrate are important for the bioreactor performance, mainly because of the anaerobic biodegradability and the composition (VS, COD and TS) of organic matter are inter-connected. Furthermore, the concentration of the substrate introduced into the reactor can also affect the metabolic activity of the microbes (Raposo *et al.*, 2011). The VS content of organic substrates is not essentially the same, because of different proteins, lipids and carbohydrates content, which represent the soluble and the easily biodegradable part. The lignin composition represents the almost non-biodegradable part of the VS. Therefore, the biodegradation and the solubility of the electron donors depend on its cellulose and lignin content which means that the hydrolysis-fermentation rate is also affected (Houbron *et al.*, 2008). The cellulose and chitin crystallinity or degree of polymerization shows different rates of degradation and this depends on the content or pretreatment done prior to their use in the methanogenesis or a sulphidogenic process. The crystallinity or degree of polymerization refers to the order of the molecules in polymer such as the α, β and γ-cellulose (Foston, 2014). Similarly, even though the COD content of electron donors with heterogeneous characteristics is different, it is also an important factor because it helps in controlling the growth rate of the SRB.

Table 2-7. Sulphate reduction efficiency reported in inverse fluidized bed bioreactors (Adapted from Reyes-Alvarado *et al.*, 2017)

Source of the inoculum	Carrier material	HRT (d)	Electron donor	COD removal efficiency (%)	Sulphate reduction efficiency (%)	COD:Sulphate ratio	References
Granular sludge from an UASB reactor treating malting process effluent (Central de Malta, Grajales, Puebla, Mexico)	Low density (267 kg.m^{-3}) polyethylene (0.4 mm diameter)	1-0.7	Mixture of VFA: acetate or lactate, propionate and butyrate	90	73	1.67-0.67	Celis-García *et al.* (2007)
Granular sludge from a UASB reactor treating paper mill wastewater (Industriewater Eerbeek B.V., Eerbeek, The Netherlands)	Low-density (400 kg.m^{-3}) polyethylene (500-1,000 μm)	2	Ethanol-lactate: 2:1-1:0 ratio	80	28	0.6	Celis *et al.* (2009)
Sulphate reducing biofilm	Low density (400 kg.m^{-3}) polyethylene (500 μm diameter)	2-1	Ethanol-lactate: 2:1-1:0 ratio	50-54	30-41	0.8	Gallegos-Garcia *et al.* (2009)
Anaerobic sludge from a digester treating activated sludge from a domestic wastewater treatment plant (De Nieuwe Waterweg in Hoek van Holland, The Netherlands)	Low density polyethylene (3 mm diameter)	1-0.37	Lactate	R1 = 14-34 and R2 = 35-68	R1 = 56-88 and R2 = 17-68	R1 = 5 and R2 = 1	Villa-Gomez *et al.* (2011)
Methanogenic granular sludge from a full scale anaerobic digester fed with buffalo manure and dairy wastewater	Polypropylene pellets (3-5 mm diameter)	1	Lactate	R1 = 35-64 and R2 = 6-61	R1 = 18-30 and R2 = 1-63	R1 = 0.67 and R2 = 0.67-4.0	Papirio *et al.* (2013a)
Anaerobic granular sludge from Biothane Systems International (Delft, The Netherlands)	Low density polyethylene beads (3 mm diameter)	0.64	Ethanol	Not reported	75	1.88	Kijjanapanich *et al.* (2014)
Anaerobic sludge from Biothane Systems International (Delft, Netherlands)	Low density polyethylene beads (3 mm diameter)	1	Ethanol	75 (pH 7.0) and 58 (pH 5.0)	74 (pH: 7.0) and 50 (pH: 5.0)	1	Janyasuthiwong *et al.* (2016)
Anaerobic sludge from a reactor digesting waste activated sludge at Harnaschpolder (The Netherlands)	Low density polyethylene beads (4 mm diameter)	1-0.5	Lactate	72-86	16-74	0.71-1.82	Reyes-Alvarado *et al.* (2017)

2.4.3.2 Particle size of electron donors

The particle size is normally considered as an important design factor because a reduction in the particle size will increase the surface area which in turn improves the performance of the biological process (De la Rubia *et al.*, 2011; Mshandete *et al.*, 2006). Sometimes, hydrolytic-fermentative bacteria find it difficult to biodegrade the organic solid waste because of its size, and therefore, it is suggested to cut or break them to allow more surface area for these microorganisms to metabolize. Since the initial hydrolysis process may take time, it is important to provide an adequate/favorable environment for the SRB to increase its metabolic activity. Failure to do so might pose a delay in the start-up or even complete failure of the bioprocess.

2.4.3.3 Source of inoculum

The adaptation of the inoculum to the bioreactor depends a lot on its origin (Behera *et al.*, 2007). Sources from thermophilic, mesophilic and halophilic conditions adapt differently when introduced into the bioreactors. For the sulphate reduction process, if the origin of the inoculum is from sulphate rich environments, it will be easier for the microorganisms to adapt themselves to the bioreactor conditions because of its sulphate content. Whereas, for sulphate deprived environments, it will take a while for the microorganisms to adapt to their new environment. But the time of adaptation depends on the syntrophic network stablished in the inoculum (Alon *et al.*, 1999; Barkai and Leibler, 1997). The inoculum from wastewater treatment plants can vary in its characteristics due to different operating conditions and daily variations, but it is mostly preferred because they all share common characteristics. The effects of the inoculum origin, concentration, activity, and storage has been reported in the literature (Raposo *et al.*, 2011). In general, start-up of sulphate reducing bioreactors could be enhanced by the introduction of inocula from sulphate rich origins. Nevertheless, IFB bioreactors are inoculated with anaerobic sludge from methanogenic reactors (Table 2-7).

2.4.3.4 Physical and chemical conditions in a bioreactor

Physical and chemical operational conditions of the reactors affect the sulphate reduction process. Physical conditions, such as volume, temperature and stirring speed have significant effects on biodegradation rate. Zagury *et al.* (2006) studied the effect of chemical conditions such as headspace gas concentrations, pH and alkalinity adjustments on the biodegradation of substrates. The volume of the reactor usually has an inverse relationship to the number of replicates that can be used and is also related to the homogeneity of the electron donor distribution. So, if the volume of the reactor is large, it reduces the amount of reproducibility of the experiment, while for substrates which are heterogeneous it would be better to use larger bioreactors. The majority of the bioreactor experiments are performed under mesophilic conditions (20 to 45°C) and a few under thermophilic conditions (45 to 60°C). Thermophilic conditions are sometimes avoided due to cost implications. Although the effect of stirring is contested, for organic solid electron donors, the stirring process will favour its contact with the SRB, increasing the microbial activity and facilitating sulphate reduction.

2.4.3.5   Biomass morphology

Different morphological features of biomass can develop during bioreactor operation. The factors that might affect the cell performance and behavior are the ones increasing stress, which at the same time can also affect the syntrophic structure.

a) *Flocs*: Flocs are a conglomeration of cells and microcolonies enmeshed in exopolymers, related to the hydrodynamics, wastewater composition, and dissolved oxygen concentrations (Dangcong *et al.*, 1999). One advantage is the fast diffusional transport compared with those in granules or biofilms (Morgan-Sagastume *et al.*, 2008). The filamentous bacteria play an important role in these structures. Mielczarek *et al.* (2012) reported that during warm periods, activated sludge flocs preserve an open structure. A high concentration of filamentous bacteria might cause settling problems.

b) *Granules*: As evolution of stress factors, the anaerobic granular sludge develops spontaneously due to auto-immobilization. This aggregation occurs in the absence of any support material, in contrast to biofilms. On the other hand, a single bacterium is not able to degrade organic matter to methane. Therefore, different bacteria, as present in granules, develop a complex and unique microbial ecosystem.

c) *Biofilms*: Planktonic cells are found in media as free floating microorganisms, but their attachment to surfaces enhances their survival in diverse environments (Rivas *et al.*, 2007). A biofilm can be defined as a complex coherent structure of cells and cellular products, like extracellular polymers (exopolysaccharide), which form and grow spontaneously attached on a static suspended solid surface (Davey and O'toole, 2000).

In the bioreactor, osmolarity, pH, oxygen, and temperature are other environmental variables that can also influence the initial biomass attachment, apart from the nature of the support material used (Ishii *et al.*, 2008). For instance, the stratification of microbial communities in biofilms can be influenced by the electron acceptors and donors (Satoh *et al.*, 2009) as well as the degradability of carbon source (Shen *et al.*, 2013). The process of biofilm formation is a multistage development, this includes attachment of microbes to the surface, cell to cell adhesion and proliferation, maturation and detachment.

2.5 Modelling biological sulphate reduction

2.5.1 Monod type modelling for biological sulphate reduction

The substrate utilization and the biokinetic constants K_s and μ_{max} can be determined using the Monod equation, given by:

$$\frac{\partial X}{\partial t} = \mu_{max} \frac{S}{K_s + S} X = \mu X \qquad \text{Eq. 2-30}$$

$$\frac{\partial S}{\partial t} = -\frac{\mu_{max}}{Y_{\frac{X}{S}}} \frac{S}{K_s + S} X = -\frac{\mu}{Y_{X/S}} X \qquad \text{Eq. 2-31}$$

Where, μ is the specific growth rate (g COD. g VSS^{-1} d^{-1}); S is the substrate concentration (g. L^{-1}); X is the biomass concentration (g VSS. L^{-1}); μ_{max} is the maximum growth rate (g COD. g VSS^{-1} d^{-1}) and K_s is the half velocity saturation coefficient (g. L^{-1}).

By increasing the initial concentration of the substrate, an increase in specific growth rate (μ) is observed until a certain concentration where the profile nearly remains stationary and reaches the μ_{max}. For every initial substrate concentration X_0, a specific bacterial growth rate (μ) or substrate consumption velocity ($v(S)$, volumetric substrate utilization rate) equation can be expressed as follows:

$$\mu = \frac{\Delta X}{\Delta t X} = v(S) \qquad \text{Eq. 2-32}$$

From Eq. 2-30 and Eq. 2-32, the following relation is obtained:

$$\mu = \mu_{max} \frac{S}{K_s + S} \qquad \text{Eq. 2-33}$$

Taking the reciprocals of both sides gives a linearized form of Eq. 2-33:

$$\frac{1}{\mu} = \frac{K_s}{\mu_{max}} + \frac{1}{\mu_{max}} \qquad \text{Eq. 2-34}$$

Eq. 2-34 gives the Lineweaver-Burke plot (Figure 2-7) for estimating the Monod constants (Ghigliazza *et al.*, 2000).

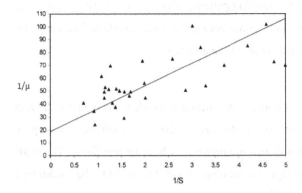

Figure 2-7. Estimation of K_S and μ_{max} for sulphate removal under steady state conditions at a COD:Sulphate ratio of 1 (Modified from Ghigliazza *et al.*, 2000)

Recently, van Wageningen *et al.* (2006) developed a kinetic model for biological sulphate reduction using primary sewage sludge as the electron donor and organic carbon source. The authors proposed a two-phase (aqueous/gas) physical, biological and chemical process based kinetic model for the methanogenic anaerobic digestion of sewage sludge. This complex model relies on the kinetic data obtained from sewage sludge hydrolysis/solubilisation and acidification, acetogenesis, and acetotrophic(clastic) and hydrogenotrophic methanogenesis. This model is still being validated with different experimental bioreactors fed with sewage sludge and sulphate, at different pH and HRT.

2.5.2 Artificial neural network (ANN) based modeling

2.5.2.1 Fundamentals of ANN

Artificial neural networks are models that can simulate a real process with great conformity inspired from functioning of biological nervous system (Nasr *et al.*, 2013). It is a mimicry of the basic structure of a biological neuron (Sodhi *et al.*, 2014). ANNs have been used in a wide array of engineering and medical applications due to their flexibility of modelling and adaptation to different case scenarios (Dragoi *et al.*, 2013; López *et al.*, 2014, 2017; Rene *et al.*, 2011a, 2011b; Reyes-Alvarado *et al.*, 2017; Soleimani *et al.*, 2013). The main aspect of these networks is their ability to learn from previous experiences and incorporate those results to make future predictions with respect to input changes (Avunduk *et al.*, 2014). ANN use the concept of artificial neurons, multi-layer perception, back propagation algorithm and mathematical functions amongst various others. The neurons of the human brain are relatively reproduced in ANNs which simulates the learning patterns of the brain. The advantage of this technology is that it is simple to apply since it does not require a mathematical function before building the model, though gives optimal results (Khataee and Kasiri, 2011).

2.5.2.2 Multi-layer perceptron

The multi-layered perceptron layer is a concept in a neural network to account for the non-linear processes taking place in the real world (Rene *et al.*, 2006). A multi-layer perception model enables neural networking to mimic real life processes which are non-linear. The multi-layer consists of an input layer, hidden layer and an output layer (Figure 2-8). The hidden layer may be composed of one or more layers. The input layer is a layer of neurons that takes signals from outside into the model, the hidden layer processes these input values within the structure of the model and the output layer gives the output of the processing (Fu *et al.*, 2013). The applied signals go through all these layers before being compared with the desired outputs and being mapped for corresponding inputs (Rady, 2011).

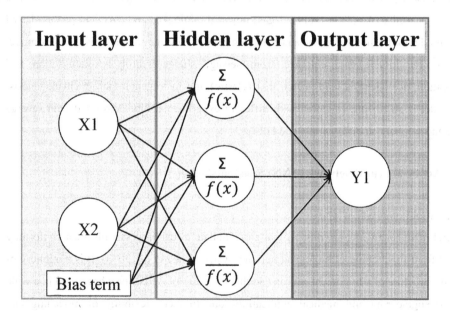

Figure 2-8. General schematic view of an artificial neural network

2.5.2.3 Back propagation algorithm

The back propagation algorithm is the most frequently used learning rule in neural networks. In fact, due to the advent of this technology, neural networks have gained interests in various fields (Avunduk *et al.*, 2014; Behera *et al.*, 2012; Rene *et al.*, 2011a, 2011b). The feed forward back propagation algorithm performs two steps: one forward step to produce a solution and a backward step to generate the error to correct the weights. The backward and forward sweeps are continuously conducted till the ANN output reaches the desired set value (Basheer and Hajmeer, 2000; Xu *et al.*, 2017). Back propagation provides an efficient method to adjust weights of the neural network, but is reported to be slow in reaching the desired output values. This technique is similar to the least mean squared error learning algorithm and is based on a descent in the gradient and the adjustment of the weights is done in a direction towards the negative gradient of the error that is measured (Hu *et al.*, 2008). Probabilistic model based different mathematical functions have been made for back propagation algorithms (Seshan *et al.*, 2014).

The perceptron algorithm can be summarized in five steps. First, to set low random values to threshold levels and weights, then to present inputs and desired output, calculate the actual output based on the formula:

$$y_k = f_h(\textstyle\sum_{i=1}^n X_i W_i) - \theta_k)$$ Eq. 2-35

Where, y_k is the solution of the model, f_h is the activation function, i is the number or signals, X_i is the input signal, W_i is the weight of each signal and θ_k is the initial threshold level.

And then set the weights based on the formula as follows:

$$w_i(new) = w_i(old) + \eta(d_k - y_k)x_i , 0 \le i \le N$$ Eq. 2-36

Where, η is the gain which is less than 1 and d_k is the desired output.

This process is repeated until the task of achieving the desired output is met (Rajasimman *et al.*, 2007).

2.5.2.4 Internal network parameters

The basic structural unit of artificial neural networks is an artificial neuron which consists of three blocks, namely thresholds, weights and activation function. The threshold levels determine the minimum amplitude of a signal to influence a process, weights determine the intensity of the influence or change and activation function defines the pattern in which a particular signal causes a disturbance or change (Basheer and Hajmeer, 2000). The nature of the activation function can be linear, binary, sigmoidal or in a few other forms. This artificial neuron network is also known as perceptron which can be trained based on some previously designed patterns (Dhussa *et al.*, 2014). The weights of the perceptron are changed based on the error from the desired output on every run of the model. Every input node, hidden node and output node is an artificial neuron. The ANN generally also involves a multi-layer network with hidden layer or layers and back propagation algorithm as learning method (Shao and Zheng, 2011).

2.5.2.5 ANN modelling for bioreactors

Wastewater quality parameters are monitored regularly by routine chemical analysis in the laboratory; however, these procedures are time consuming and labor intensive. In order to monitor and control process variables in highly fluctuating industrial wastewater treatment systems, an adaptive process control device is required. Under such conditions, an online ANN based software sensor integrated with process control system would be preferable for monitoring, predicting and controlling fluctuating state variables. ANN based control and prediction systems have been tested for several bioreactor configurations, such as sequencing batch reactor (SBR), continuous stirred tank bioreactor (CSTR), fluidized bed reactor (FBR), inverse fluidized bed (IFB) reactor and activated sludge process (ASP). Hong *et al.* (2007) operated a SBR in 8 h cyclic mode and each cycle consisted of 2 h anaerobic, 4 h 30 min aerobic, 1 h 30 min settling and drawing phase. The solids retention time (SRT) was maintained by periodically wasting mixed liquor suspended solids (MLSS) at the end of the aerobic stage. Multiway principal component analysis (MPCA) was used for analyzing the whole process data consisting of 24 batches. MPCA was used to compress the normal batch data and extract the information by projecting the data onto a low dimensional space that summarizes both the variables and their time trajectories. Easily available online measurements such as pH, ORP and DO were taken as inputs, while the nutrient concentrations, *e.g.* NH_4^+, NO_3^- and PO_4^{3-}, were considered as the outputs for the ANN model.

ANN modelling has also been applied for predicting sulphate removal in different biological wastewater treatment systems. Atasoy *et al.* (2013) showed that the predicted results from the neural network model fitted the actual experimental data well and the results could be used to reduce operational costs and risks. Using a feed-forward ANN to train a genetic algorithm, Vinod *et al.* (2009) simulated the degradation process of phenol by Pseudomonas sp. in a fluidized bed reactor (FBR). The model proposed by Sahinkaya *et al.* (2007) used a three layered network topology consisting of 20 neurons in the hidden layer to predict the effluent sulfate, acetate, sulfide or alkalinity concentrations using easily measurable process parameters as the inputs. In that study, the Levenberg-Marquardt algorithm was chosen after considering several other algorithms. The results from that study suggest that the control of the operational conditions can be carried out based on the predictions done by ANN models for enhanced performance of the FBR. Sahinkaya (2009) modelled the sulfidogenic treatment of sulfate and zinc containing simulated wastewater in a mesophilic CSTR. The author used a two-layer ANN with a tan-sigmoid transfer function for the hidden layer and a linear transfer function for the output layer. Feed pH, sulfate, Zn, COD and operation time were used as input parameters to predict the effluent sulfate, COD, acetate and Zn concentrations.

Wang *et al.* (2009) modelled a complex denitrifying sulfide removal (DSR) process that has complex interactions between autotrophic and heterotrophic denitrifiers using ANN. The steady-state performance of an expended granular sludge bed (EGSB)-DSR bioreactor for nitrite denitrification and the complete DSR process was predicted using a four-layered network topology, comprising of two hidden layers. The results showed that the DSR performance was affected by the process parameters in the order: HRT > influent sulfide concentration > C/S ratio > N/S ratio. Using a three layered network topology (3-7-3), the standard back-propagation training algorithm with gradient descent and a sensitivity analysis, Janyasuthiwong *et al.* (2016) modelled the COD and sulfate removal efficiencies as well as the total sulfide production profiles in the IFB reactor using a three-layered ANN. Based on sensitivity analysis, the pH was determined to be the most important parameter affecting sulphate reduction in an IFB bioreactor. Reyes-Alvarado *et al.* (2017) used a three layered network topology (5-11-3, Figure 2-9) to model IFB performance, and reported that the sulphate reduction efficiency is hampered if the COD:sulphate ratio is the limiting factor in an IFB bioreactor. Limiting COD:sulphate ratio can affect the production of sulphide and carbonate; therefore, the buffering capacity as well as the pH decreases. These results clearly showed that the influent sulphate concentration and the pH are crucial parameters that affect process intensification of an IFB bioreactor.

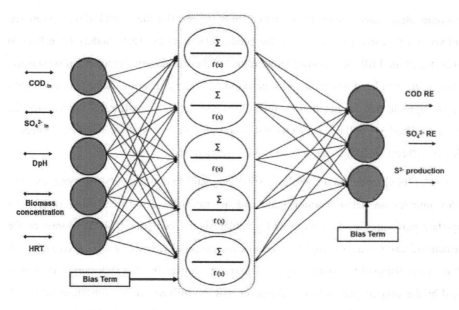

Figure 2-9. A three-layered ANN model used to predict the COD and sulphate removal efficiency in an inverse fluidized bed bioreactor (Adapted from Reyes-Alvarado *et al.*, 2017)

2.6 Conclusions

The variations in concentrations of sulphate and carbon source, pH and the presence of competing ions affect biochemical activities in wastewater treatment systems. The use of SRB technology for sulphate rich wastewater treatment is advantageous due to minimal sludge production, ability to perform simultaneous oxidation of organic matter and the reduction of sulfate. Sulphate reduction in bioreactors is affected by parameters such as the type of electron donor, COD:Sulphate ratio, pH, HRT and reactor configuration. Sulphate reduction efficiencies > 90% is achieved when COD is not a limiting factor in a bioreactor, and a COD:Sulphate ratio > 2.0 is recommended in such cases. Most of the sulphate reducing bioreactors are also able to handle fluctuations in COD or sulphate loading rates. The ability of the SRB to overcome feast and famine periods clearly shows the application of this technology for industrial situations.

2.7 References

Al-Zuhair, S., El-Naas, M.H., Al-Hassani, H., 2008. Sulfate inhibition effect on sulfate reducing bacteria. J. Biochem. Technol. 1, 39–44.

Alon, U., Surette, M.G., Barkai, N., Leibler, S., 1999. Robustness in bacterial chemotaxis. Nature 397, 168–171. doi:10.1038/16483

Andalib, M., Zhu, J., Nakhla, G., 2012. A new definition of bed expansion index and voidage for fluidized biofilm-coated particles. Chem. Eng. J. 189–190, 244–249. doi:10.1016/j.cej.2012.02.065

Arne Alphenaar, P., Visser, A., Lettinga, G., 1993. The effect of liquid upward velocity and hydraulic retention time on granulation in UASB reactors treating wastewater with a high sulphate content. Bioresour. Technol. 43, 249–258. doi:10.1016/0960-8524(93)90038-D

Atasoy, A.D., Babar, B., Sahinkaya, E., 2013. Artificial neural network prediction of the performance of upflow and downflow fluidized bed reactors treating acidic mine drainage water. Mine Water Environ. 32, 222–228. doi:10.1007/s10230-013-0232-x

Avunduk, E., Tumac, D., Atalay, A.K., 2014. Prediction of roadheader performance by artificial neural network. Tunn. Undergr. Sp. Technol. 44, 3–9. doi:10.1016/j.tust.2014.07.003

Baena, S., Fardeau, M.L., Labat, M., Ollivier, B., Garcia, J.L., Patel, B.K., 1998. *Desulfovibrio aminophilus sp. nov.*, a novel amino acid degrading and sulfate reducing bacterium from an anaerobic dairy wastewater lagoon. Syst. Appl. Microbiol. 21, 498–504. doi:10.1016/S0723-2020(98)80061-1

Barkai, N., Leibler, S., 1997. Robustness in simple biochemical networks. Nature 387, 913–917. doi:10.1038/43199

Barton, L.L., Fauque, G.D., 2009. Biochemistry, physiology and biotechnology of sulfate-reducing bacteria, in: Advances in Applied Microbiology. Elsevier Inc., pp. 41–98. doi:10.1016/S0065-2164(09)01202-7

Basheer, I.., Hajmeer, M., 2000. Artificial neural networks: fundamentals, computing, design and application. J. Microbiol. Methods 43, 3–31. doi:10.1016/S0167-7012(00)00201-3

Baskaran, V., Nemati, M., 2006. Anaerobic reduction of sulfate in immobilized cell bioreactors, using a microbial culture originated from an oil reservoir. Biochem. Eng. J. 31, 148–159. doi:10.1016/j.bej.2006.07.007

Bertolino, S.M., Rodrigues, I.C.B., Guerra-Sá, R., Aquino, S.F., Leão, V.A., 2012. Implications of volatile fatty acid profile on the metabolic pathway during continuous sulfate reduction. J. Environ. Manage. 103, 15–23. doi:10.1016/j.jenvman.2012.02.022

Butlin, K.R., Selwyn, S.C., Wakerley, D.S., 1956. Sulphide production from sulphate-enriched sewage sludges. J. Appl. Bacteriol. 19, 3–15. doi:10.1111/j.1365-2672.1956.tb00036.x

Celis-García, L.B., Razo-Flores, E., Monroy, O., 2007. Performance of a down-flow fluidized bed reactor under sulfate reduction conditions using volatile fatty acids as electron donors. Biotechnol. Bioeng. 97, 771–779. doi:10.1002/bit.21288

Celis, L.B., Villa-Gómez, D., Alpuche-Solís, A.G., Ortega-Morales, B.O., Razo-Flores, E., 2009. Characterization of sulfate-reducing bacteria dominated surface communities during start-up of a down-flow fluidized bed reactor. J. Ind. Microbiol. Biotechnol. 36, 111–121. doi:10.1007/s10295-008-0478-7

Chen, Y., Cheng, J.J., Creamer, K.S., 2008. Inhibition of anaerobic digestion process: a review. Bioresour. Technol. 99, 4044–64. doi:10.1016/j.biortech.2007.01.057

Cirik, K., Dursun, N., Sahinkaya, E., Çinar, Ö., 2013. Effect of electron donor source on the treatment of Cr(VI)-containing textile wastewater using sulfate-reducing fluidized bed reactors (FBRs). Bioresour. Technol. 133, 414–420. doi:10.1016/j.biortech.2013.01.064

D´Acunto, B., Esposito, G., Frunzo, L., Pirozzi, F., 2011. Dynamic modeling of sulfate reducing biofilms. Comput. Math. with Appl. 62, 2601–2608. doi:10.1016/j.camwa.2011.07.064

Dangcong, P., Bernet, N., Delgenes, J., Moletta, R., 1999. Aerobic granular sludge: a case report. Water Res. 33, 890–893. doi:10.1016/S0043-1354(98)00443-6

Davey, M.E., O'toole, G.A., 2000. Microbial biofilms: from ecology to molecular genetics. Microbiol. Mol. Biol. Rev. 64, 847–867. doi:10.1128/MMBR.64.4.847-867.2000

De la Rubia, M.A., Fernández-Cegrí, V., Raposo, F., Borja, R., 2011. Influence of particle size and chemical composition on the performance and kinetics of anaerobic digestion process of sunflower oil cake in batch mode. Biochem. Eng. J. 58–59, 162–167. doi:10.1016/j.bej.2011.09.010

Dhussa, A.K., Sambi, S.S., Kumar, S., Kumar, S., Kumar, S., 2014. Nonlinear autoregressive exogenous modeling of a large anaerobic digester producing biogas from cattle waste. Bioresour. Technol. 170, 342–349. doi:10.1016/j.biortech.2014.07.078

Diez Blanco, V., García Encina, P.A., Fdz-Polanco, F., 1995. Effects of biofilm growth, gas and liquid velocities on the expansion of an anaerobic fluidized bed reactor (AFBR). Water Res. 29, 1649–1654. doi:10.1016/0043-1354(95)00001-2

Dragoi, E.-N., Curteanu, S., Galaction, A.-I., Cascaval, D., 2013. Optimization methodology based on neural networks and self-adaptive differential evolution algorithm applied to an aerobic fermentation process. Appl. Soft Comput. 13, 222–238.

doi:10.1016/j.asoc.2012.08.004

du Preez, L.A., Odendaal, J.P., Maree, J.P., Ponsonby, M., 1992. Biological removal of sulphate from industrial effluents using producer gas as energy source. Environ. Technol. 13, 875–882. doi:10.1080/09593339209385222

Dvorak, D.H., Hedin, R.S., Edenborn, H.M., McIntire, P.E., 1992. Treatment of metal contaminated water using bacterial sulfate reduction: results from pilot-scale reactors. Biotechnol. Bioeng. 40, 609–616.

Eldyasti, A., Nakhla, G., Zhu, J., 2012. Influence of particles properties on biofilm structure and energy consumption in denitrifying fluidized bed bioreactors (DFBBRs). Bioresour. Technol. 126, 162–171. doi:10.1016/j.biortech.2012.07.113

Escudié, R., Cresson, R., Delgenès, J.-P., Bernet, N., 2011. Control of start-up and operation of anaerobic biofilm reactors: an overview of 15 years of research. Water Res. 45, 1–10. doi:10.1016/j.watres.2010.07.081

Fontes Lima, D.M., Zaiat, M., 2012. The influence of the degree of back-mixing on hydrogen production in an anaerobic fixed-bed reactor. Int. J. Hydrogen Energy 37, 9630–9635. doi:10.1016/j.ijhydene.2012.03.097

Foston, M., 2014. Advances in solid-state NMR of cellulose. Curr. Opin. Biotechnol. 27, 176–184. doi:10.1016/j.copbio.2014.02.002

Fu, Z.-J., Xie, W.-F., Luo, W.-D., 2013. Robust on-line nonlinear systems identification using multilayer dynamic neural networks with two-time scales. Neurocomputing 113, 16–26. doi:10.1016/j.neucom.2012.11.041

Gallegos-Garcia, M., Celis, L.B., Rangel-Méndez, R., Razo-Flores, E., 2009. Precipitation and recovery of metal sulfides from metal containing acidic wastewater in a sulfidogenic down-flow fluidized bed reactor. Biotechnol. Bioeng. 102, 91–99. doi:10.1002/bit.22049

Genari, A.N., Passos, F.V., Passos, F.M.L., 2003. Configuration of a bioreactor for milk lactose hydrolysis. J. Dairy Sci. 86, 2783–2789. doi:10.3168/jds.S0022-0302(03)73875-2

Ghigliazza, R., Lodi, A., Rovatti, M., 2000. Kinetic and process considerations on biological reduction of soluble and scarcely soluble sulfates. Resour. Conserv. Recycl. 29, 181–194. doi:10.1016/S0921-3449(99)00055-5

Gibson, G.R., 1990. Physiology and ecology of the sulphate-reducing bacteria. J. Appl.

Bacteriol. 69, 769–797. doi:10.1111/j.1365-2672.1990.tb01575.x

Goršek, A., Tramšek, M., 2008. Kefir grains production-An approach for volume optimization of two-stage bioreactor system. Biochem. Eng. J. 42, 153–158. doi:10.1016/j.bej.2008.06.009

Grein, F., Ramos, A.R., Venceslau, S.S., Pereira, I.A.C., 2013. Unifying concepts in anaerobic respiration: insights from dissimilatory sulfur metabolism. Biochim. Biophys. Acta - Bioenerg. 1827, 145–160. doi:10.1016/j.bbabio.2012.09.001

Hansen, T.A., 1993. Carbon metabolism of sulfate-reducing bacteria, in: Odom, J.M., Singleton, J.R. (Eds.), The Sulfate-Reducing Bacteria: Contemporary Perspectives. Springer New York, New York, NY, pp. 21–40. doi:10.1007/978-1-4613-9263-7_2

Hao, O.J., Chen, J.M., Huang, L., Buglass, R.L., 1996. Sulfate-reducing bacteria. Crit. Rev. Environ. Sci. Technol. 26, 155–187. doi:10.1080/10643389609388489

Hong, S.H., Lee, M.W., Lee, D.S., Park, J.M., 2007. Monitoring of sequencing batch reactor for nitrogen and phosphorus removal using neural networks. Biochem. Eng. J. 35, 365–370. doi:10.1016/j.bej.2007.01.033

Hoover, R., 2001. Composition, molecular structure, and physicochemical properties of tuber and root starches: a review. Carbohydr. Polym. 45, 253–267. doi:10.1016/S0144-8617(00)00260-5

Houbron, E., González-López, G.I., Cano-Lozano, V., Rustrían, E., 2008. Hydraulic retention time impact of treated recirculated leachate on the hydrolytic kinetic rate of coffee pulp in an acidogenic reactor. Water Sci. Technol. 58, 1415–1421. doi:10.2166/wst.2008.492

Hu, D., Liu, H., Yang, C., Hu, E., 2008. The design and optimization for light-algae bioreactor controller based on artificial neural network-model predictive control. Acta Astronaut. 63, 1067–1075. doi:10.1016/j.actaastro.2008.02.008

Hulshoff Pol, L.W., De Castro Lopes, S.I., Lettinga, G., Lens, P.N.L., 2004. Anaerobic sludge granulation. Water Res. 38, 1376–1389. doi:10.1016/j.watres.2003.12.002

Ishii, S., Shimoyama, T., Hotta, Y., Watanabe, K., 2008. Characterization of a filamentous biofilm community established in a cellulose-fed microbial fuel cell. BMC Microbiol. 8, 1–12. doi:10.1186/1471-2180-8-6

Janyasuthiwong, S., Rene, E.R., Esposito, G., Lens, P.N.L., 2016. Effect of pH on the performance of sulfate and thiosulfate-fed sulfate reducing inverse fluidized bed reactors. J. Environ. Eng. 142, 1–11. doi:10.1061/(ASCE)EE.1943-7870.0001004

Jeihanipour, A., Niklasson, C., Taherzadeh, M.J., 2011. Enhancement of solubilization rate of cellulose in anaerobic digestion and its drawbacks. Process Biochem. 46, 1509–1514. doi:10.1016/j.procbio.2011.04.003

Jing, Z., Hu, Y., Niu, Q., Liu, Y., Li, Y.-Y., Wang, X.C., 2013. UASB performance and electron competition between methane-producing archaea and sulfate-reducing bacteria in treating sulfate-rich wastewater containing ethanol and acetate. Bioresour. Technol. 137, 349–357. doi:10.1016/j.biortech.2013.03.137

John, R.P., Nampoothiri, K.M., Pandey, A., 2007. Fermentative production of lactic acid from biomass: an overview on process developments and future perspectives. Appl. Microbiol. Biotechnol. 74, 524–534. doi:10.1007/s00253-006-0779-6

Kaksonen, A.H., Plumb, J.J., Robertson, W.J., Riekkola-Vanhanen, M., Franzmann, P.D., Puhakka, J.A., 2006. The performance, kinetics and microbiology of sulfidogenic fluidized-bed treatment of acidic metal- and sulfate-containing wastewater. Hydrometallurgy 83, 204–213. doi:10.1016/j.hydromet.2006.03.025

Kang, H., Weiland, P., 1993. Ultimate anaerobic biodegradability of some agro-industrial residues. Bioresour. Technol. 43, 107–111. doi:10.1016/0960-8524(93)90168-B

Kelleher, B.P., Leahy, J.J., Henihan, A.M., O'Dwyer, T.F., Sutton, D., Leahy, M.J., 2002. Advances in poultry litter disposal technology - a review. Bioresour. Technol. 83, 27–36. doi:10.1016/S0960-8524(01)00133-X

Khataee, A.R., Kasiri, M.B., 2011. Modeling of biological water and wastewater treatment processes using artificial neural networks. Clean - Soil, Air, Water 39, 742–749. doi:10.1002/clen.201000234

Kieu, H.T.Q., Müller, E., Horn, H., 2011. Heavy metal removal in anaerobic semi-continuous stirred tank reactors by a consortium of sulfate-reducing bacteria. Water Res. 45, 3863–3870. doi:10.1016/j.watres.2011.04.043

Kijjanapanich, P., Do, A.T., Annachhatre, A.P., Esposito, G., Yeh, D.H., Lens, P.N.L., 2014. Biological sulfate removal from construction and demolition debris leachate: effect of bioreactor configuration. J. Hazard. Mater. 269, 38–44. doi:10.1016/j.jhazmat.2013.10.015

Leustek, T., Martin, M.N., Bick, J.-A., Davies, J.P., 2000. Pathways and regulation of sulphur metabolism revealed through molecular and genetic studies. Annu. Rev. Plant Physiol. Plant Mol. Biol. 51, 141–165. doi:10.1146/annurev.arplant.51.1.141

Li, Y., Park, S.Y., Zhu, J., 2011. Solid-state anaerobic digestion for methane production from organic waste. Renew. Sustain. Energy Rev. 15, 821–826. doi:10.1016/j.rser.2010.07.042

Liamleam, W., Annachhatre, A.P., 2007. Electron donors for biological sulfate reduction. Biotechnol. Adv. 25, 452–463. doi:10.1016/j.biotechadv.2007.05.002

Lier, J.B., Zee, F.P., Frijters, C.T.M.J., Ersahin, M.E., 2015. Celebrating 40 years anaerobic sludge bed reactors for industrial wastewater treatment. Rev. Environ. Sci. Bio/Technology 14, 681–702. doi:10.1007/s11157-015-9375-5

Maree, J.P., Hill, E., 1989. Biological removal of sulphate from industrial effluents and concomitant production of sulphur. Water Sci. Technol. 21, 265–276.

Mata-Alvarez, J., Macé, S., Llabrés, P., 2000. Anaerobic digestion of organic solid wastes. An overview of research achievements and perspectives. Bioresour. Technol. 74, 3–16. doi:10.1016/S0960-8524(00)00023-7

Matias, P.M., Pereira, I.A.C., Soares, C.M., Carrondo, M.A., 2005. Sulphate respiration from hydrogen in bacteria: a structural biology overview. Prog. Biophys. Mol. Biol. 89, 292–329. doi:10.1016/j.pbiomolbio.2004.11.003

Mielczarek, A.T., Kragelund, C., Eriksen, P.S., Nielsen, P.H., 2012. Population dynamics of filamentous bacteria in Danish wastewater treatment plants with nutrient removal. Water Res. 46, 3781–3795. doi:10.1016/j.watres.2012.04.009

Mizuno, O., Li, Y.Y., Noike, T., 1998. The behavior of sulfate-reducing bacteria in acidogenic phase of anaerobic digestion. Water Res. 32, 1626–1634. doi:10.1016/S0043-1354(97)00372-2

Moestedt, J., Nilsson Påledal, S., Schnürer, A., 2013. The effect of substrate and operational parameters on the abundance of sulphate-reducing bacteria in industrial anaerobic biogas digesters. Bioresour. Technol. 132, 327–332. doi:10.1016/j.biortech.2013.01.043

Molino, A., Nanna, F., Ding, Y., Bikson, B., Braccio, G., 2013. Biomethane production by anaerobic digestion of organic waste. Fuel 103, 1003–1009. doi:10.1016/j.fuel.2012.07.070

Morgan-Sagastume, F., Larsen, P., Nielsen, J.L., Nielsen, P.H., 2008. Characterization of the loosely attached fraction of activated sludge bacteria. Water Res. 42, 843–854. doi:10.1016/j.watres.2007.08.026

Mshandete, A., Björnsson, L., Kivaisi, A.K., Rubindamayugi, M.S.T., Mattiasson, B., 2006. Effect of particle size on biogas yield from sisal fibre waste. Renew. Energy 31, 2385–2392. doi:10.1016/j.renene.2005.10.015

Muyzer, G., Stams, A.J.M., 2008. The ecology and biotechnology of sulphate-reducing bacteria. Nat. Rev. Microbiol. 6, 441–454. doi:10.1038/nrmicro1892

Nasr, N., Hafez, H., El Naggar, M.H., Nakhla, G., 2013. Application of artificial neural networks for modeling of biohydrogen production. Int. J. Hydrogen Energy 38, 3189–3195. doi:10.1016/j.ijhydene.2012.12.109

Nedwell, D.B., Reynolds, P.J., 1996. Treatment of landfill leachate by methanogenic and sulphate-reducing digestion. Water Res. 30, 21–28. doi:10.1016/0043-1354(95)00128-8

O'Sullivan, C., Burrell, P.C., Clarke, W.P., Blackall, L.L., 2008. The effect of biomass density on cellulose solubilisation rates. Bioresour. Technol. 99, 4723–4731. doi:10.1016/j.biortech.2007.09.070

Örlygsson, J., Houwen, F.P., Svensson, B.H., 1994. Influence of hydrogenothrophic methane formation on the thermophilic anaerobic degradation of protein and amino acids. FEMS Microbiol. Ecol. 13, 327–334. doi:10.1111/j.1574-6941.1994.tb00079.x

Ozmihci, S., Kargi, F., 2008. Ethanol production from cheese whey powder solution in a packed column bioreactor at different hydraulic residence times. Biochem. Eng. J. 42, 180–185. doi:10.1016/j.bej.2008.06.017

Papirio, S., Esposito, G., Pirozzi, F., 2013a. Biological inverse fluidized-bed reactors for the treatment of low pH- and sulphate-containing wastewaters under different COD conditions. Environ. Technol. 34, 1141–1149. doi:10.1080/09593330.2012.737864

Papirio, S., Villa-Gomez, D.K., Esposito, G., Pirozzi, F., Lens, P.N.L., 2013b. Acid mine drainage treatment in fluidized-bed bioreactors by sulfate-reducing bacteria: a critical review. Crit. Rev. Environ. Sci. Technol. 43, 2545–2580. doi:10.1080/10643389.2012.694328

Peixoto, G., Saavedra, N.K., Varesche, M.B.A., Zaiat, M., 2011. Hydrogen production from soft-drink wastewater in an upflow anaerobic packed-bed reactor. Int. J. Hydrogen Energy

36, 8953–8966. doi:10.1016/j.ijhydene.2011.05.014

Pepper, I.L., Rensing, C., Gerba, C.P., 2004. Environmental microbial properties and processes, in: Environmental Monitoring and Characterization. Elsevier (USA), pp. 263–280. doi:10.1016/B978-012064477-3/50016-3

Pillai, C.K.S., Paul, W., Sharma, C.P., 2009. Chitin and chitosan polymers: chemistry, solubility and fiber formation. Prog. Polym. Sci. 34, 641–678. doi:10.1016/j.progpolymsci.2009.04.001

Rabus, R., Hansen, T.A., Widdel, F., 2006. Dissimilatory Sulfate- and Sulfur-Reducing Prokaryotes, in: The Prokaryotes. Springer New York, New York, NY, pp. 659–768. doi:10.1007/0-387-30742-7_22

Rady, H.A.K., 2011. Shannon entropy and mean square errors for speeding the convergence of multilayer neural networks: a comparative approach. Egypt. Informatics J. 12, 197–209. doi:10.1016/j.eij.2011.09.002

Rajasimman, M., Govindarajan, I., Karthikeyan, C., 2007. Artificial neural network modeling of an inverse fluidized bed bioreactor. J. Appl. Sci. Environ. Manag. 11, 65–69. doi:10.4314/jasem.v11i2.54991

Raposo, F., Fernández-Cegrí, V., de la Rubia, M.A., Borja, R., Béline, F., Cavinato, C., Demirer, G., Fernández, B., Fernández-Polanco, M., Frigon, J.C., Ganesh, R., Kaparaju, P., Koubova, J., Méndez, R., Menin, G., Peene, A., Scherer, P., Torrijos, M., Uellendahl, H., Wierinck, I., de Wilde, V., 2011. Biochemical methane potential (BMP) of solid organic substrates: evaluation of anaerobic biodegradability using data from an international interlaboratory study. J. Chem. Technol. Biotechnol. 86, 1088–1098. doi:10.1002/jctb.2622

Rasi, S., Veijanen, A., Rintala, J., 2007. Trace compounds of biogas from different biogas production plants. Energy 32, 1375–1380. doi:10.1016/j.energy.2006.10.018

Rene, E.R., Estefanía López, M., Veiga, M.C., Kennes, C., 2011a. Neural network models for biological waste-gas treatment systems. N. Biotechnol. 29, 56–73. doi:10.1016/j.nbt.2011.07.001

Rene, E.R., López, M.E., Veiga, M.C., Kennes, C., 2011b. Artificial neural network modelling for waste: gas and wastewater treatment applications, in: Computational Modeling and Simulation of Intellect. IGI Global, Hershey PA, USA, pp. 224–263. doi:10.4018/978-1-

60960-551-3.ch010

Reyes-Alvarado, L.C., Okpalanze, N.N., Kankanala, D., Rene, E.R., Esposito, G., Lens, P.N.L., 2017. Forecasting the effect of feast and famine conditions on biological sulphate reduction in an anaerobic inverse fluidized bed reactor using artificial neural networks. Process Biochem. 55, 146–161. doi:10.1016/j.procbio.2017.01.021

Rinaudo, M., 2006. Chitin and chitosan: properties and applications. Prog. Polym. Sci. 31, 603–632. doi:10.1016/j.progpolymsci.2006.06.001

Rivas, L., Dykes, G.A., Fegan, N., 2007. A comparative study of biofilm formation by Shiga toxigenic *Escherichia coli* using epifluorescence microscopy on stainless steel and a microtitre plate method. J. Microbiol. Methods 69, 44–51. doi:10.1016/j.mimet.2006.11.014

Robertson, L.A., Kuenen, J.G., 2006. The Colorless Sulfur Bacteria, in: The Prokaryotes. Springer New York, New York, NY, pp. 985–1011. doi:10.1007/0-387-30742-7_31

Sahinkaya, E., Özkaya, B., Kaksonen, A.H., Puhakka, J.A., 2007. Neural network prediction of thermophilic (65°C) sulfidogenic fluidized-bed reactor performance for the treatment of metal-containing wastewater. Biotechnol. Bioeng. 97, 780–787. doi:10.1002/bit.21282

Sampaio, R.M.M., Timmers, R.A., Kocks, N., André, V., Duarte, M.T., van Hullebusch, E.D., Farges, F., Lens, P.N.L., 2010. Zn–Ni sulfide selective precipitation: the role of supersaturation. Sep. Purif. Technol. 74, 108–118. doi:10.1016/j.seppur.2010.05.013

Sarti, A., Zaiat, M., 2011. Anaerobic treatment of sulfate-rich wastewater in an anaerobic sequential batch reactor (AnSBR) using butanol as the carbon source. J. Environ. Manage. 92, 1537–1541. doi:10.1016/j.jenvman.2011.01.009

Satoh, H., Odagiri, M., Ito, T., Okabe, S., 2009. Microbial community structures and *in situ* sulfate-reducing and sulfur-oxidizing activities in biofilms developed on mortar specimens in a corroded sewer system. Water Res. 43, 4729–4739. doi:10.1016/j.watres.2009.07.035

Seshan, H., Goyal, M.K., Falk, M.W., Wuertz, S., 2014. Support vector regression model of wastewater bioreactor performance using microbial community diversity indices: effect of stress and bioaugmentation. Water Res. 53, 282–296. doi:10.1016/j.watres.2014.01.015

Shao, H., Zheng, G., 2011. Convergence analysis of a back-propagation algorithm with adaptive momentum. Neurocomputing 74, 749–752. doi:10.1016/j.neucom.2010.10.008

Shen, Z., Zhou, Y., Wang, J., 2013. Comparison of denitrification performance and microbial diversity using starch/polylactic acid blends and ethanol as electron donor for nitrate removal. Bioresour. Technol. 131, 33–39. doi:10.1016/j.biortech.2012.12.169

Sheoran, A.S., Sheoran, V., Choudhary, R.P., 2010. Bioremediation of acid-rock drainage by sulphate-reducing prokaryotes: a review. Miner. Eng. 23, 1073–1100. doi:10.1016/j.mineng.2010.07.001

Show, K.Y., Lee, D.J., Chang, J.S., 2011. Bioreactor and process design for biohydrogen production. Bioresour. Technol. 102, 8524–8533. doi:10.1016/j.biortech.2011.04.055

Sodhi, S.S., Chandra, P., Tanwar, S., 2014. A new weight initialization method for sigmoidal feedforward artificial neural networks, in: 2014 International Joint Conference on Neural Networks (IJCNN). IEEE, pp. 291–298. doi:10.1109/IJCNN.2014.6889373

Soleimani, R., Shoushtari, N.A., Mirza, B., Salahi, A., 2013. Experimental investigation, modeling and optimization of membrane separation using artificial neural network and multi-objective optimization using genetic algorithm. Chem. Eng. Res. Des. 91, 883–903. doi:10.1016/j.cherd.2012.08.004

Stams, A.J.M., Plugge, C.M., de Bok, F.A.M., van Houten, B.H.G.W., Lens, P., Dijkman, H., Weijma, J., 2005. Metabolic interactions in methanogenic and sulfate-reducing bioreactors. Water Sci. Technol. 52, 13–20.

Taguchi, Y., Sugishima, M., Fukuyama, K., 2004. Crystal structure of a novel zinc-binding ATP Sulfurylase from *Thermus thermophilus* HB8. Biochemistry 43, 4111–4118. doi:10.1021/bi036052t

Tang, K., Baskaran, V., Nemati, M., 2009. Bacteria of the sulphur cycle: an overview of microbiology, biokinetics and their role in petroleum and mining industries. Biochem. Eng. J. 44, 73–94. doi:10.1016/j.bej.2008.12.011

Tsukamoto, T.K., Killion, H.A., Miller, G.C., 2004. Column experiments for microbiological treatment of acid mine drainage: Low-temperature, low-pH and matrix investigations. Water Res. 38, 1405–1418. doi:10.1016/j.watres.2003.12.012

van Haandel, A., Kato, M.T., Cavalcanti, P.F.F., Florencio, L., 2006. Anaerobic reactor design concepts for the treatment of domestic wastewater. Rev. Environ. Sci. Bio/Technology 5, 21–38. doi:10.1007/s11157-005-4888-y

van Houten, R.T., van der Spoel, H., van Aelst, A.C., Hulshoff Pol, L.W., Lettinga, G., 1996.

Biological sulfate reduction using synthesis gas as energy and carbon source. Biotechnol. Bioeng. 50, 136–144. doi:10.1002/(SICI)1097-0290(19960420)50:2<136::AID-BIT3>3.0.CO;2-N

Venu Vinod, A., Arun Kumar, K., Venkat Reddy, G., 2009. Simulation of biodegradation process in a fluidized bed bioreactor using genetic algorithm trained feedforward neural network. Biochem. Eng. J. 46, 12–20. doi:10.1016/j.bej.2009.04.006

Villa-Gomez, D., Ababneh, H., Papirio, S., Rousseau, D.P.L., Lens, P.N.L., 2011. Effect of sulfide concentration on the location of the metal precipitates in inversed fluidized bed reactors. J. Hazard. Mater. 192, 200–207. doi:10.1016/j.jhazmat.2011.05.002

Villa-Gomez, D.K., Cassidy, J., Keesman, K.J., Sampaio, R., Lens, P.N.L., 2014. Sulfide response analysis for sulfide control using a pS electrode in sulfate reducing bioreactors. Water Res. 50, 48–58. doi:10.1016/j.watres.2013.10.006

Ward, A.J., Hobbs, P.J., Holliman, P.J., Jones, D.L., 2008. Optimisation of the anaerobic digestion of agricultural resources. Bioresour. Technol. 99, 7928–7940. doi:10.1016/j.biortech.2008.02.044

Weijma, J., Stams, A.J.M., Hulshoff Pol, L.W., Lettinga, G., 2000. Thermophilic sulfate reduction and methanogenesis with methanol in a high rate anaerobic reactor. Biotechnol. Bioeng. 67, 354–363. doi:10.1002/(SICI)1097-0290(20000205)67:3<354::AID-BIT12>3.0.CO;2-X

Widdel, F., 2006. The Genus *Desulfotomaculum*, in: The Prokaryotes. Springer US, New York, NY, pp. 787–794. doi:10.1007/0-387-30744-3_25

Xia, Y., Cai, L., Zhang, T., Fang, H.H.P., 2012. Effects of substrate loading and co-substrates on thermophilic anaerobic conversion of microcrystalline cellulose and microbial communities revealed using high-throughput sequencing. Int. J. Hydrogen Energy 37, 13652–13659. doi:10.1016/j.ijhydene.2012.02.079

Yang, Y., Tsukahara, K., Yagishita, T., Sawayama, S., 2004. Performance of a fixed-bed reactor packed with carbon felt during anaerobic digestion of cellulose. Bioresour. Technol. 94, 197–201. doi:10.1016/j.biortech.2003.11.025

Zagury, G.J., Kulnieks, V.I., Neculita, C.M., 2006. Characterization and reactivity assessment of organic substrates for sulphate-reducing bacteria in acid mine drainage treatment. Chemosphere 64, 944–954. doi:10.1016/j.chemosphere.2006.01.001

Zhou, J., He, Q., Hemme, C.L., Mukhopadhyay, A., Hillesland, K., Zhou, A., He, Z., Van Nostrand, J.D., Hazen, T.C., Stahl, D.A., Wall, J.D., Arkin, A.P., 2011. How sulphate-reducing microorganisms cope with stress: lessons from systems biology. Nat. Rev. Microbiol. 9, 452–466. doi:10.1038/nrmicro2575

Chapter 3

Forecasting the effect of feast and famine conditions on biological sulphate reduction in an anaerobic inverse fluidized bed reactor using artificial neural networks

A modified version of this chapter was published as:
Reyes-Alvarado L.C., Okpalanze N.N., Kankanala D., Rene E.R., Esposito G., Lens P.N.L., (2017). Forecasting the effect of feast and famine conditions on biological sulphate reduction in an anaerobic inverse fluidized bed reactor using artificial neural networks. Process Biochem. 55, 146–161. doi:10.1016/j.procbio.2017.01.021

Abstract

The longevity and robustness of bioreactors used for wastewater treatment is determined by the activity of the microorganisms under steady and transient loading conditions. Two identical continuously operated inverse fluidized bed bioreactors (IFB), IFB R1 and IFB R2, were tested for sulphate removal under the same operating conditions for 140 d (periods I-IV). Later, IFB R1 was used as the control reactor (period V), while IFB R2 was operated under feast (period V-A) and famine (period V-B) feeding conditions for 66 d. The sulphate removal efficiency was comparable in both IFB, < 20% in period I and ~ 70% during periods II, III and IV. The robustness of the IFB was evident when the sulphate removal efficiency remained comparable during the feast period (67 ± 15%) applied to IFB R2 compared to continuous feeding periods (period IV (71 ± 4%) for IFB R2 and period V (61 ± 15%) for IFB R1). The IFB performance was modelled using a three-layered artificial neural networks (ANN) model (5-11-3) and a sensitivity analysis, the sulphate removal was found to be dependent on the COD:sulphate ratio. Besides, the robustness, resilience and adaptation time of the IFB were affected by the degree of mixing and the hydraulic retention time.

Keywords: Inverse fluidized bed reactor; biological sulphate reduction; feast-famine conditions; artificial neural networks; sensitivity analysis

3.1 Introduction

Industrial wastewaters rich in sulphate (0.4-20.8 g $SO_4^{2-}.L^{-1}$) are characterised by a low pH, high redox potential and low to negligible chemical oxygen demand (COD) concentrations, but contain high concentrations of heavy metals in the case of acid mine drainage (AMD) [1–3]. As a consequence, this type of wastewater can dramatically damage the flora and fauna of water reservoirs, rivers, groundwater and land when they are not treated efficiently [4]. Biological sulphate reduction under non controlled conditions results in the release of toxic sulphide from these wastewaters into the environment, but under controlled conditions, the sulphide can be used to precipitate heavy metals [5].

The use of anaerobic reactors for the treatment of sulphate rich wastewater has been widely reported in the literature [6,7]. The lack of COD in many industrial wastewaters, including AMD, makes biological removal of sulphate by sulphate reducing bacteria (SRB) in anaerobic reactors an expensive process, because of the use of industrial grade chemicals as electron donors. To minimize the electron donor supply and to maintain high removal efficiencies of the pollutants during long term operation, the reactors hydrodynamics as well as the key process parameters should be optimized. Many types of anaerobic reactor configurations have been tested and applied for the biological removal of sulphate: the batch reactor [8], sequencing batch reactor [9], upflow anaerobic sludge blanket reactor, (UASB) [10], extended granular sludge bed reactor (EGSB) [6], fixed bed reactor [11], fluidized bed reactor [7] and gas lift reactor [12]. In all these reactor configurations, the effects of the COD:sulphate ratio, the use of different electron donors, temperature (mesophilic and thermophilic conditions) and hydraulic residence time (HRT) have been tested.

Table 3-1. Sulphate reduction efficiency reported in inverse fluidized bed bioreactors

Source of the inoculum	Carrier material	HRT (d)	Electron donor	COD removal efficiency (%)	Sulphate removal efficiency (%)	COD:Sulphate ratio	References
Granular sludge from an UASB reactor treating malting process effluent (Central de Malta, Grajales, Puebla, Mexico)	Low density (267 kg.m^{-3}) polyethylene (0.4 mm diameter)	1-0.7	Mixture of VFA: acetate or lactate, propionate and butyrate	90	73	1.67-0.67	Celis-García et al. [13]
Granular sludge from a UASB reactor treating paper mill wastewater (Industriewater Eerbeek B.V., Eerbeek, The Netherlands)	Low-density (400 kg.m^{-3}) polyethylene (500-1,000 µm)	2	Ethanol-lactate: 2:1-1:0 ratio	80	28	0.6	Celis et al. [14]
Sulphate reducing biofilm	Low density (400 kg.m^{-3}) polyethylene (500 µm diameter)	2-1	Ethanol-lactate: 2:1-1:0 ratio	50-54	30-41	0.8	Gallegos-Garcia et al. [15]
Anaerobic sludge from a digester treating activated sludge from a domestic wastewater treatment plant (De Nieuwe Waterweg in Hoek van Holland, The Netherlands)	Low density polyethylene (3 mm diameter)	1-0.37	Lactate	R1 = 14-34 and R2 = 35-68	R1 = 56-88 and R2 = 17-68	R1 = 5 R2 = 1	Villa-Gomez et al. [16]
Methanogenic granular sludge from a full scale anaerobic digester fed with buffalo manure and dairy wastewater	Polypropylene pellets (3-5 mm diameter)	1	Lactate	R1 = 35-64 and R2 = 6-61	R1 = 18-30 and R2 = 1-63	R1 = 0.67 R2 = 0.67-4.0	Papirio et al. [17]
Anaerobic granular sludge from Biothane Systems International (Delft, The Netherlands)	Low density polyethylene beads (3 mm diameter)	0.64	Ethanol	Not reported	75	1.88	Kijjanapanich et al. [7]
Anaerobic sludge from Biothane Systems International (Delft, Netherlands)	Low density polyethylene beads (3 mm diameter)	1	Ethanol	75 (pH 7.0) and 58 (pH 5.0)	74 (pH: 7.0) and 50 (pH: 5.0)	1	Janyasuthiwong et al. [18]
Anaerobic sludge from a reactor digesting waste activated sludge at Harnaschpolder (The Netherlands)	Low density polyethylene beads (4 mm diameter)	1-0.5	Lactate	72-86	16-74	0.71-1.82	This research

Inverse fluidized bed reactors (IFB) have also been studied for biological sulphate reduction. The pH and COD:sulphate ratio are the major factors affecting the IFB reactor performance (Table 3-1). For instance, in IFB reactors the feasibility of using volatile fatty acids (VFA) as electron donors for the SRB [13] and microbial community characterization [14] have been studied previously. Furthermore, the use of the IFB reactor is an innovative process intensification strategy to achieve biological sulphate reduction, increase biomass retention, and improve the metal-sulphide precipitation for the treatment of wastewater from AMD [15,16]. The effect of pH on the sulphate reduction process was demonstrated and the authors showed that a decrease in the pH of the influent from 7.0 to 3.0 leads to a complete failure of the reactor operation [17]. Nevertheless, when the influent pH was increased to 5.0, the sulphate removal efficiency increased to 97% at a COD:sulphate ratio of 4.0. In another study, when the pH of a sulphidogenic IFB reactor was lowered to 5.0, the sulphate removal efficiency dropped to < 40% from ~ 75% at pH 7.0 [18].

The implementation and use of artificial neural networks (ANN) to model wastewater and waste gas treatment systems is growing exponentially. Besides their applications in the field of waste treatment, they have also been successfully tested to develop multistep ahead predictive models that can predict heat load of consumers attached to district heating systems, friction factors in pipes by coupling ANN to adaptive neuro-fuzzy inference system (ANFIS), and future solar radiation based on a series of meteorological data [19–22]. In a recent study, under steady state feeding conditions, artificial neural networks (ANN) modelling was successfully tested in an IFB reactor and it was possible to map the COD removal efficiency, sulphate removal efficiency and the sulphide production with a network topology of 3-7-3 (input-hidden-output layer, respectively) [18]. Besides, the authors also performed a sensitivity analysis on the input parameters and showed that the sulphidogenic process was mainly affected by the influent pH [18].

To our knowledge, the robustness of sulphate reduction in an IFB under transient feeding conditions has not yet been reported, neither the application of ANN to elucidate the robustness, performance and relationship of the process variables under such feeding conditions. Thus, this research aimed at investigating the effect of transient feeding conditions on the biological removal of sulphate in two continuously operated IFB reactors. The hydrodynamics of the IFB bioreactor were evaluated by performing residence time distribution (RTD) studies. The ANN was used to predict the efficiency of biological sulphate reduction using hydraulic retention time (HRT), biomass concentration, influent sulphate and COD concentration as well as the difference in pH (DpH) values between the effluent and influent as the input parameters.

3.2   Material and methods

3.2.1   Synthetic wastewater composition

The synthetic wastewater used for performing experiments in the IFB had the following composition (mg.L^{-1}): NH$_4$Cl (300), MgCl$_2$·6H$_2$O (120), KH$_2$PO$_4$ (200), KCl (250), CaCl$_2$·2H$_2$O (15), yeast extract (20) and 0.5 mL of a mixture of micronutrients [16]. The composition of trace elements was prepared with FeCl$_2$·4H$_2$O (1500), MnCl$_2$·4H$_2$O (100), EDTA (500), H$_3$BO$_3$ (62), ZnCl$_2$ (70), NaMoO$_4$·2H$_2$O (36), AlCl$_3$·6H$_2$O (40), NiCl$_3$·6H$_2$O (24), CoCl$_2$·6H$_2$O (70), CuCl$_2$·2H$_2$O (20) and HCl 36 % (1 mL.L^{-1}). Sodium lactate was also used as the electron donor and sodium sulphate was used as the electron acceptor. All reagents used in this study were of analytical grade.

3.2.2   Carrier material

The carrier material used in the IFB reactors was low density polyethylene beads (Sigma Aldrich) with a diameter of 4 mm. Prior to their use, the beads were rinsed with demineralized water in order to remove the smaller fractions. 0.3 L of this carrier material was used in both IFB reactors.

3.2.3   Inoculum

The inoculum was obtained from a municipal wastewater treatment plant, located at Harnaschpolder (The Netherlands). The consortia of bacteria were taken from the anaerobic reactor digesting waste activated sludge. The seed sludge for the two IFB bioreactors (R1 and R2) contained 30,812 (± 2,096) mg.L^{-1} of total suspended solids (TSS) and 20,364 (± 1,535) mg.L^{-1} of volatile suspended solids (VSS), respectively.

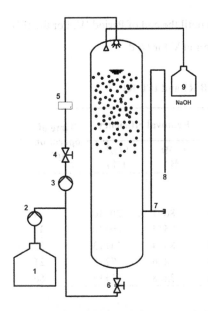

Figure 3-1. Schematic of the inverse fluidized bed reactor
Main components: 1) influent tank, 2) peristaltic pump, 3) recirculation pump, 4) recirculation control valve, 5) rotameter (only used during hydrodynamic evaluation), 6) safety valve, 7) sampling area, 8) effluent connected to the sewage pipe and 9) gas trap

3.2.4   Anaerobic IFB bioreactor set up

The schematic of the IFB reactors used in this study is shown in Figure 3-1. The reactors were built with transparent polyvinyl chloride (PVC) tubes (internal diameter 0.056 m and height 1.03 m). The effective volume of the reactor was 2.46 L, corresponding to 1 m of reactor height. The influent was supplied to the system with the help of a peristaltic pump (Masterflex L/S) and it was connected to the recirculation tubing and placed before the recirculation pump (Iwaki Magnet Pump, Iwaki Co. Ltd. Tokyo, Japan). The liquid inside the IFB bioreactor was recirculated at a constant velocity of 300 L.h^{-1}, resulting in a down flow velocity of 122 m.h^{-1}. The effluent port was placed at a height of 0.2 m from the bottom of each reactor. The effluent pipe was placed upwards in parallel to the IFB bioreactor and folded close to the top of the bioreactor in order to maintain a constant liquid level inside the IFB bioreactor.

3.2.5   IFB bioreactor operational conditions

Two IFB bioreactors were started simultaneously and named as IFB R1 (control reactor) and IFB R2 (operated first in steady and later under feast and famine conditions). The operational conditions are given in Table 3-2. Both reactors were operated in continuous mode from day 0 and the reactors followed the same operational schedule until the end of Period IV. For the IFB R1, the steady feeding conditions were maintained in Period V for 66 d.

Table 3-2. Operational conditions and performance of IFB R1 and IFB R2

Period of operation	COD_{in} (mg. L^{-1})	$SO_4^{2-}{}_{in}$ (mg. L^{-1})	HRT (d)	COD/SO_4^{2-} ratio	Removal efficiency COD (%)	Removal efficiency SO_4^{2-} (%)	Time of operation (d)
				IFB R1			
I	530	746	1	0.71	80±7	20±10	34
II	1,358	746	1	1.82	75±5	69±10	32
III	1,000	746	1	1.34	86±4	74±13	52
IV	1,000	746	0.5	1.34	78±6	72±5	21
V	1,000	746	0.5	1.34	72±8	61±15	66
				IFB R2			
I	530	746	1	0.71	65±16	16±14	34
II	1,358	746	1	1.82	67±13	70±10	32
III	1,000	746	1	1.34	84±6	68±9	52
IV	1,000	746	0.5	1.34	75±6	71±4	21
V-A (feast)	2,133	1,495	0.5	1.43	78±7	67±15	66
V-B (famine)	0	0	0.5	0.00	-70±282	-552±928	

Note: The performance of both IFB bioreactors in terms of COD and sulphate removal efficiencies (RE)

The experimental conditions for IFB R2 in Period V (66 d) were split into V-A and V-B. In Period V-A, the IFB R2 was fed with both COD and sulphate for 3 days (feast) and subsequently followed by 4 days of operation in the absence of both COD and sulphate (famine condition, Period V-B). The electron donor and electron acceptor concentrations were doubled during the feast conditions (COD = 2,133 mg.L^{-1} and SO_4^{2-} = 1,495 mg.L^{-1}), while the HRT was maintained constant at 0.5 d. The famine conditions refer to the omission of the electron donor and acceptor from the influent; all other components were kept in the influent. These transient or feast and famine conditions were applied 10 times in succession. The performance of IFB R2 was compared to that of IFB R1 as indicated in section 3.2.8.1.

3.2.6 RTD studies

In the IFB reactor, prior to the start of the sulphate reduction experiments, RTD experiments were carried out by injecting a spike (2 mL) of sodium chloride (1 M) as tracer at the point of entry of the influent in the recirculation pipe. MiliQ water was used as the eluent, while the conductivity of this water was used as the base line for performing RTD studies. Conductivity of the IFB bioreactor effluent was measured and converted into concentration units with the help of a calibration plot. A constant flow rate of 5 L.d^{-1} or an HRT of 0.5 d was maintained during these experiments. The experimental data was processed as described in section 3.2.8.2

3.2.7 Chemical analysis

All chemical analysis that were performed in this study as well as the biomass concentration estimations (TSS and VSS) were done according to the procedures outlined in Standard Methods for the Examination of Water and Wastewater [23]. Briefly, COD concentrations were determined by the close reflux colorimetric method, total dissolved sulphide (S^{2-}) by the methylene blue reaction method and sulphate by ion chromatography (ICS-1000 Dionex, ASI-100 Dionex) fitted with a IonPac AS14n column.

3.2.8 Data processing

3.2.8.1 Performance and comparison of the IFB bioreactors

The concentration data of the influent and effluent of both IFB reactors were monitored periodically. The differences in the sulphate reduction performance were evaluated and compared by plotting the removal rate (*RR*) in the "*y*" axis against the loading rate (*LR*) in the "*x*" axis for both IFB R1 and R2. *LR* and *RR* were defined as follows:

$$LR = (Q/V_R) \times (A_{in}) \hspace{4cm} \text{Eq. 3-1}$$

$$RR = (Q/V_R) \times (A_{in} - B_{out}) \hspace{3.5cm} \text{Eq. 3-2}$$

The removal efficiencies (*RE*) were calculated using the following equation:

$$RE = (A_{in} - B_{out}/A_{in}) \times 100 \hspace{3.5cm} \text{Eq. 3-3}$$

Where A_{in} is the influent concentration (mg.L^{-1}) of COD or sulphate, Bout is the effluent concentration (mg.L^{-1}) of COD or sulphate, Q is the flow rate (L.d^{-1}) and V_R (L) is the IFB reactor volume. Q was set to 2.5 L.d^{-1} or 5 L.d^{-1} based on the period of IFB operation, while the V_R was 2.5 L. The biomass concentration present on the surface of the carrier materials was obtained using a correlation between biomass concentration (mg.L^{-1}) and VSS (mg.L^{-1}), as reported in the literature [24]. According to Eq. 3-4, 83.4% of the biomass attached to the carrier material was biofilm, while the rest was suspended biomass (16.6%):

$$Biomass = (VSS \times 0.834)/0.166 \qquad \text{Eq. 3-4}$$

3.2.8.2 Evaluation of RTD

The time series data of concentration were processed using Microsoft Excel. The RTD function $E(t)$ was defined as follows:

$$E(t) = C(t)/\int_0^\infty C(t)\,dt \qquad \text{Eq. 3-5}$$

and obtained when dividing each concentration data "$C(t)$" by the area below the curve of the profile of $C(t)$ against time ($\int_0^\infty C(t)\,dt$). $E(t)$ has units of time (h^{-1}).

The addition of each $E(t)$ data evaluated at different times corresponded to the cumulative profile, $F(t)$:

$$F(t) = \int_0^t E(t)\,dt \qquad \text{Eq. 3-6}$$

The mean residence time was evaluated using the following equation:

$$\int_0^\infty tE(t)\,dt = t_m \qquad \text{Eq. 3-7}$$

Then, the RTD data was normalized as follows:

$$E(\theta) = E(t) \times t_m \qquad \text{Eq. 3-8}$$

Likewise, the normalized time (θ) was defined by:

$$\theta = t/t_m \qquad \text{Eq. 3-9}$$

The profiles $E(\theta)$ against θ and $F(t)$ against θ were plotted. More details concerning the RTD procedure and data analysis can be found in [25].

3.2.8.3 ANN modelling

It is important to select the input parameters that are likely to have a major impact on the process and the desired output for better interpretation of the ANN model results. As shown in Figure 3-2, the process of developing the best network architecture involves a number of network specifications to be optimized: the number of neurons in the hidden layer, the learning rate, epoch size, momentum, the processing element activation function and the training count of the network. This more detailed optimization procedure concerning these network parameters has been described in detail elsewhere [26–28]. After obtaining the desired network topology and to extract more process based information from the model, the strength of the relationship between the output variable and input variable was estimated by performing a sensitivity analysis.

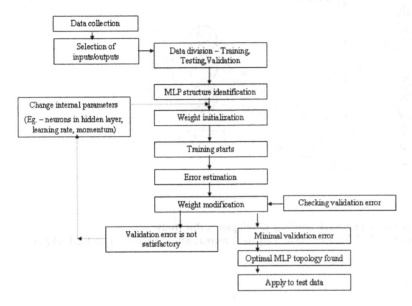

Figure 3-2. Flow chart illustrating the different steps involved in developing a neural network model

The data from the different phases of IFB R2 operation (206 d) was used for ANN modelling. It consists of 62 data points for different process variables monitored in this study. The input variables for the ANN model were HRT, biomass concentration in the reactor, sulphate and COD concentration in the influent and the difference in pH values between the effluent and the influent (DpH). The output of the ANN model was the COD and sulphate removal efficiencies and sulphide concentrations. The general network topology developed in this study is shown in Figure 3-3. The data was divided as training (~70%) and testing (~30 %) set; thus, 42 data points were used for training the network, while the remaining 20 data points were used to test the developed model. The basic statistics of the training and test data are shown in

Table 3-3 and Table 3-4, respectively. The data was normalized and scaled to the range 0 to 1 in order to suit the transfer function in the hidden (sigmoid) and output (linear) layer. The following equation was used to normalize the data:

$$\hat{X} = (X - X_{min})/(X_{max} - X_{min})$$ Eq. 3-10

Where, $\hat{X}$ is the normalized value, X_{min} and X_{max} are the minimum and maximum values of X, respectively.

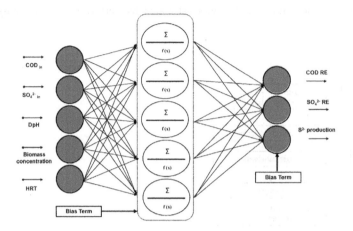

Figure 3-3. Three layered network topology developed in this study
Network architecture = 5-11-3 consisting of 5 inputs, 3 outputs and 11 neurons in the hidden layer

Table 3-3. Basic statistics of the training data

Variable	N	Mean	Standard deviation	Minimum	Maximum
HRT (d)	42	0.714	0.250	0.500	1.000
COD (mg. L^{-1})	42	1096.667	263.319	275.000	1675.000
SO$_4^{2-}$ (mg. L^{-1})	42	787.499	181.428	427.950	1424.280
Biomass (mg. L^{-1})	42	1442.633	1571.706	100.480	7435.660
DpH	42	1.630	0.146	1.350	2.060
COD-RE (%)	42	73.456	13.691	44.170	93.710
SO$_4^{2-}$ -RE (%)	42	62.399	15.643	17.010	82.580
Sulphide (mg. L^{-1})	42	143.413	55.534	18.130	298.130

Note: N is the number of data points used for training the ANN model

Table 3-4. Basic statistics of the test data

Variable	N	Mean	Standard deviation	Minimum	Maximum
HRT (d)	20	0.800	0.251	0.500	1.000
COD (mg. L^{-1})	20	1029.000	203.509	741.670	1491.670
SO_4^{2-} (mg. L^{-1})	20	771.785	40.097	724.500	879.980
Biomass (mg. L^{-1})	20	629.267	593.649	100.480	2461.810
DpH	20	1.611	0.156	1.420	1.930
COD-RE (%)	20	77.795	11.053	44.670	90.340
SO_4^{2-}-RE (%)	20	62.750	13.433	30.110	88.310
Sulphide (mg. L^{-1})	20	118.657	55.307	12.500	189.380

Note: N is the number of data points used for testing the ANN model

The different internal parameters of the network: the number of neurons in the hidden layer (N_H), learning rate (η), momentum term (μ), training count (T_C) and training vector size (ε) were chosen based on a trial and error approach. This was done by performing simulations at different initial settings of these parameters, *i.e.*, by keeping one or two parameters at their default values, and varying the other parameters slowly from low to high values. During the network training, the value of ε was kept constant at 40 as the number of data points used for training was 42. The best values of these network parameters obtained after the model training are shown in Table 3-5.

Table 3-5. Best values of network parameters used to train the ANN model developed for an IFB reactor

Training parameters	Parameter values	
	Range of values tested	Best value
Training count (T_C)	20000 to 70000	62382
Number of neurons in input layer (N_I)	5	5
Number of neurons in hidden layer (N_H)	5 to 12	11
Number of neurons in output layer (N_O)	3	3
Learning rate (η)	0.2 to 1	0.9
Momentum term (μ)	0.1 to 1	0.8
Error tolerance	0.0001	
Number of training data set (NT_r)	42	
Number of test data set (NT_e)	20	

The closeness of prediction between the experimental and the ANN model fitted outputs were evaluated by computing the determination coefficient values, as shown by:

$$R^2 = \left\{ \left[\sum_{i=1}^{N} (Y_{model\,i} - \overline{Y_{model}})(Y_{observed\,i} - \overline{Y_{observed}}) \right] \middle/ \left[(N-1) S_{Y_{model}} S_{Y_{observed}} \right] \right\}^2$$

Eq. 3-11

Where, R^2 is the determination coefficient, $Y_{model\,i}$ the model predictions, $Y_{observed\,i}$ the observed true values from experiments, N the number of cases analyzed, $\overline{Y}$ the average value and S_Y the standard deviations.

A sensitivity analysis was carried out on the different input variables used in ANN modelling. The absolute average sensitivity (AAS) was calculated by the addition of the changes in the output variables caused by moving the input variables by a small amount over the entire training set. This can be defined as follows:

$$S_{ki,abs} = \left(\sum_{p=1}^{p} \left| S_{ki}^{(p)} \right| \right) / p$$

Eq. 3-12

Where, $S_{ki}^{(p)}$ is the sensitivity of a trained output and p is the number of training patterns.

The connection weights between the different neurons in the input, hidden and output layers were computed using the following equation:

$$I_j = \sum_{m=1}^{m=N_h} \left[\left(|W_{jm}^{ih}| \middle/ \sum_{k=1}^{N_i} |W_{km}^{ih}| \right) \times |W_{mn}^{ho}| \right] \middle/ \sum_{k=1}^{k=N_i} \left\{ \sum_{m=1}^{m=N_h} \left[\left(|W_{km}^{ih}| \middle/ \sum_{k=1}^{N_i} |W_{km}^{ih}| \right) \times |W_{mn}^{ho}| \right] \right\}$$

Eq. 3-13

Where, N_i and N_h are the number of neurons in the input and hidden neurons, respectively, W is the connection weight, superscripts 'i', 'h', 'o' denote the input, hidden and output layers, respectively, and subscripts 'k', 'm' and 'n' refer to the input, hidden and output neurons, respectively.

In this study, the back error propagation (BEP) training algorithm was used, this is a generalization of the delta rule to multi-layered feedforward networks developed by Rumelhart et al. [28]. This algorithm minimizes the error function with respect to the connection weights between the input-hidden and hidden-output layers. The global error function E for a particular set of training vector samples can be estimated as follows:

$$E = 1/2 \sum (O_d - O_p)^2 \qquad\qquad \text{Eq. 3-14}$$

Where, E is the global error function, O_d is the desired output and O_p is the output predicted by the network.

ANN modelling was carried out using the shareware version of the multivariable statistical modelling software, NNMODEL (PA, USA). More detailed information on the application and development of neural models for biological wastewater treatment has been described elsewhere [26,29].

3.3 Results

3.3.1 RTD of the IFB bioreactor

The IFB bioreactor showed a mixing time value of $\theta = 0.151$, as seen from the RTD profiles (Figure 3-4). This was equivalent to a time of 0.0755 d (1.81 h) that was calculated using Eq. 3-15:

$$t = \tau_m \times \theta \qquad\qquad \text{Eq. 3-15}$$

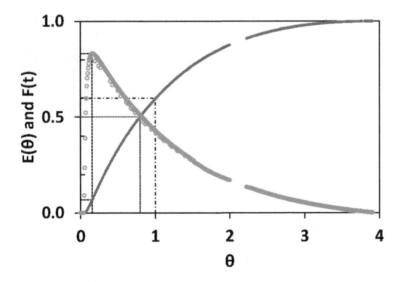

Figure 3-4. The residence time distribution of an IFB reactor tested at HRT=12 h
The dimensionless profiles corresponding to $E(\theta)$ ($\circ$) and the cumulative curve of the mass $F(t)$ (+) leaving the IFB reactor were plotted against θ. The mixing time $\theta = 0.151$ (---) was obtained when $E(\theta)$ = 0.83 and $F(t) = 0.069$. The fraction of the mass equivalent to $F(t) = 0.5$ leaving the reactor (....) was observed at $\theta = 0.797$. The mean residence time was $\tau_m = 12.008$ h ($\theta = 1$) and the fraction of mass which left the reactor (-.-.-) was $F(t) = 0.598$

This time also corresponded to the highest NaCl concentration leaving the reactor, $E(\theta) = 0.83$, and is equivalent to the NaCl fraction $F(t) = 0.069$ (6.9%) inside the IFB bioreactor. The NaCl fraction equivalent to 50% ($F(t) = 0.5$) left the reactor at $\theta = 0.797$ (0.398 d or 9.57 h). The mean residence time was calculated as $\tau_m = 12.008$ h (≈ 0.5 d) that also corresponds to a θ value of 1. At a HRT of 12.008 h (≈ 0.5 d), 59.8 % of NaCl mass left the reactor, $i.e.$ $F(t) = 0.598$. Based on the RTD results, it can be ascertained that almost 4 times the HRT ($\theta = 3.9$) was required to completely wash out NaCl from the reactor.

3.3.2 Biological sulphate reduction under steady state feeding conditions

Two IFB bioreactors, $viz.$ IFB R1 and IFB R2, were operated in continuous mode under the same operating conditions and schedules. The first 140 d of operation corresponded to the periods of operation I, II, III and IV as shown in Figure 3-5 and Figure 3-6, respectively. During this time, changes were applied only when steady state response was reached in the two IFB, apart from period I to II where the COD was increased from ~ 530 to ~ 1,350 mg.L^{-1} to ensure sulphate reduction.

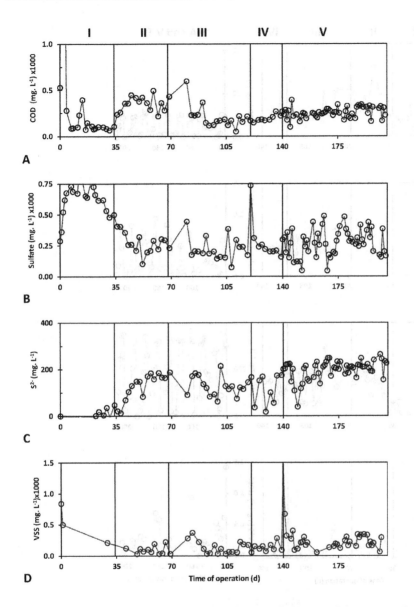

Figure 3-5. Performance of IFB R1 operated at steady feeding conditions (control reactor)
Effluent concentration profiles of COD (A), sulphate (B), sulphide (C) and VSS (D)

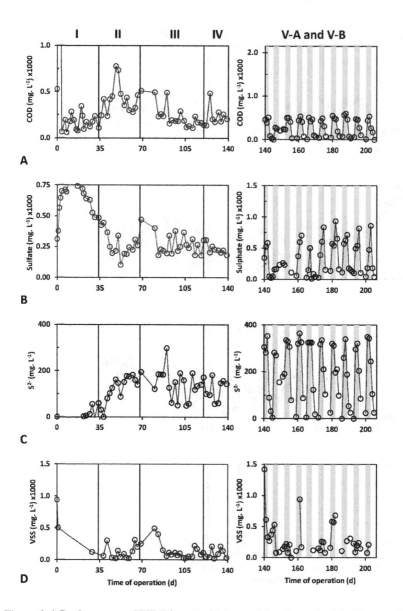

Figure 3-6. Performance of IFB R2 operated at steady and transient feeding conditions
IFB R2 was operated at steady (periods of operation I, II, III and IV) and transient [V-A (feast 3 d, shaded bars) and V-B (famine 4 d, colorless bars)] feeding conditions. Effluent concentration profiles of COD (A), sulphate (B), sulphide (C) and VSS (D)

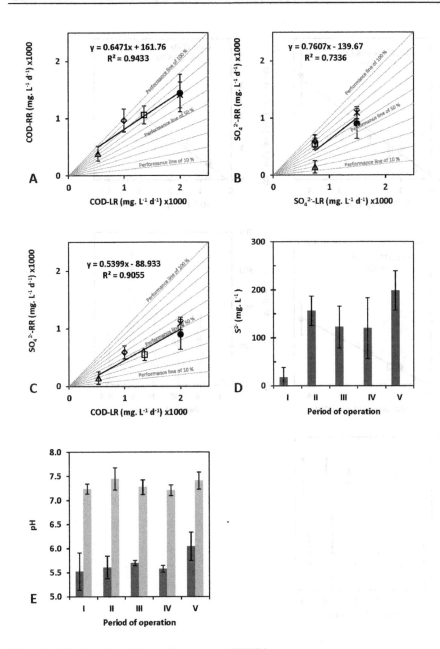

Figure 3-7. Evaluation of the performance of IFB R1
The periods of operation are represented as follows: I ($\triangle$), II ($\square$), III ($\diamondsuit$), IV ($*$) and V ($\bullet$). A) The COD removal rate (COD-*RR*) was compared to the COD loading rate (COD-*LR*). B) The sulphate removal rate (SO$_4^{2-}$-*RR*) was compared to the sulphate loading rate (SO$_4^{2-}$-*LR*). C) The sulphate removal rate (SO$_4^{2-}$-*RR*) was compared to the COD loading rate (COD-*LR*). D) The sulphide produced due to sulphate reduction in the different periods. E) The pH in the influent (dark bars) and the pH in the effluent (clear bars) during each period of operation

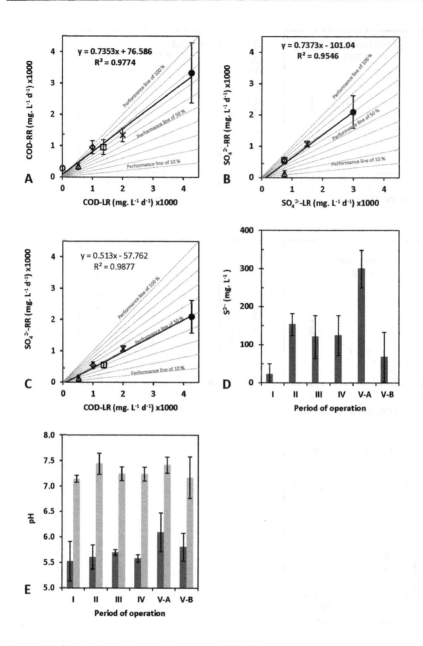

Figure 3-8. Evaluation of the performance of IFB R2

The periods of operation are represented as follows: I ($\triangle$), II ($\square$), III ($\diamond$), IV (*), the feast conditions V-A ($\bullet$) and the famine conditions V-B (o). A) The COD removal rate (COD-RR) was compared to the COD loading rate (COD-LR). B) The sulphate removal rate (SO$_4^{2-}$-RR) was compared to the sulphate loading rate (SO$_4^{2-}$-LR). C) The sulphate removal rate (SO$_4^{2-}$-RR) was compared to the COD loading rate (COD-LR). D) The sulphide produced due to sulphate reduction in the different periods. E) The pH in the influent (dark bars) and the pH in the effluent (clear bars) during each period of operation

During period I, at an HRT of 1 d and a COD:sulphate ratio of 0.71, the COD removal efficiency was 80 ($\pm$ 7)% and 65 ($\pm$ 16)% for IFB R1 and IFB R2, respectively. Besides, the sulphate removal efficiency was 20 ($\pm$ 10)% and 16 ($\pm$ 14)% in the two IFB reactors. The COD was increased in period II, irrespective of the sulphate concentration in the effluent from the previous period. In this period, at a HRT of 1 d and a COD:sulphate ratio of 1.82, the COD removal efficiency was maintained at nearly similar values of 75 ($\pm$ 5)% for IFB R1 and 67 ($\pm$ 13)% for IFB R2, respectively. The sulphate removal efficiency was positively influenced, *i.e.* it achieved 69 ($\pm$ 10)% and 70 ($\pm$ 10)% for IFB R1 and IFB R2, respectively. Thereafter, in the third period, at an HRT of 1 d, the COD concentration was adjusted to reach a COD:sulphate ratio of 1.34. Under these conditions, the COD removal efficiency was 86 ($\pm$ 4)% and 84 ($\pm$ 6)%, while the sulphate removal efficiency was 74 ($\pm$ 13)% and 68 ($\pm$ 9)% for IFB R1 and IFB R2, respectively.

In period IV, the HRT was changed to 0.5 d, while the COD:sulphate ratio was kept constant at 1.34. Among the two IFB, IFB R1 showed a decline in the sulphate removal efficiency by ~ 10% on the first day due to this HRT change; however, sulphate removal was restored after one day of operation. The COD removal efficiency was 78 ($\pm$ 6)% and 75 ($\pm$ 6)%, while the sulphate removal efficiency was 72 ($\pm$ 5)% and 71 ($\pm$ 4)% for IFB R1 and IFB R2, respectively. These changes in the performance are shown in the profiles depicted in c and Figure 3-6.

Concerning biomass concentrations, during the first period of operation, wash out of VSS from the reactors (Figure 3-5D and Figure 3-6D) was noticed. During start up, the initial VSS concentrations were > 800 and < 1,000 mg VSS.L^{-1} for IFB R1 and IFB R2, respectively, and these values decreased to ~ 150 mg VSS.L^{-1}. Nevertheless, the VSS concentration reached a nearly steady value after the first period of operation, ~ 250 mg VSS.L^{-1} in both IFB.

During the IFB operation, the sulphide production increased during the first two periods of operation. After 50 days of operation, both reactors showed a sulphide production at concentrations higher than 100 mg.L^{-1} for the periods of operation II, III and IV (Figure 3-5C and Figure 3-6C). The sulphide production is shown in Figure 3-7D and Figure 3-8D for the IFB R1 and IFB R2, respectively.

3.3.3 Biological sulphate reduction under non steady feeding conditions

Non steady or transient feeding experiments were carried out from days 140 to 206, *i.e.* for 66 d. The reference IFB bioreactor (IFB R1, Figure 3-5 – Period V) operated continuously under steady feeding conditions (HRT = 0.5, COD = 1,000 mg.L^{-1} and SO$_4^{2-}$ = 746 mg.L^{-1}) and was compared to IFB R2 that was fed discontinuously to affect the feast and famine or transient conditions (Figure 3-6, periods V-A and V-B).

The IFB R1 showed a COD removal efficiency of 72 ($\pm$ 8)% and a sulphate removal efficiency of 61 ($\pm$ 15)% during period V. During the feast period (V-A), the IFB R2 evidenced a COD removal efficiency of 78 ($\pm$ 7)% and a sulphate removal efficiency of 67 ($\pm$ 15)%. During the famine period (V-B), the IFB R2 showed negative values for the COD and sulphate removal efficiencies, -70 ($\pm$ 282)% and -552 ($\pm$ 928)%, respectively. Concerning the sulphide production in the IFB, the sulphide concentration reached $\sim$ 200 mg.L^{-1} in IFB R1 during period V of operation (Figure 3-5C). The IFB R2 produced $\sim$ 300 mg.L^{-1} of sulphide during the feast period (V-A). During famine conditions (V-B), the effluent concentration of sulphide dropped to zero (Figure 3-6C). The VSS concentration in both IFB bioreactors remained at $\sim$ 250 mg.L^{-1} (Figure 3-5D and Figure 3-6D).

3.3.4 ANN Modelling

3.3.4.1 Selecting the best training network parameters

The relationship between the input variables (COD$_{in}$, SO$_4^{2-}$$_{in}$, DpH, biomass concentration and HRT) and the output variables (COD RE, SO$_4^{2-}$ RE and the S^{2-} concentration produced) in biological sulphate reduction was studied and the data from IFB R2 was used for developing the ANN model. Based on this, the number of neurons in the input (N_i = 5) and output (N_o = 3) layers were assigned to the ANN model. The training count (T_c) and number of neurons (N_h) in the hidden layer were identified to determine the suitable network topology. The T_c was varied from 20,000 to 70,000 by keeping other network parameters such as learning rate (η = 0.5) and momentum term (μ = 0.5) at their constant values. The best value for T_c was found to be 62,382. N_h was varied from 5 to 12 and the best value was found to be 11. The best values of η (0.9) and μ (0.8) were determined by a trial and error approach, by varying these parameters between narrow intervals of 0.1. A high correlation was found between the experimental data and model fitted values when these settings were used (Table 3-5). The R^2 values were 0.87, 0.77 and 0.72 for COD removal efficiency, sulphate removal efficiency and sulphide production, respectively.

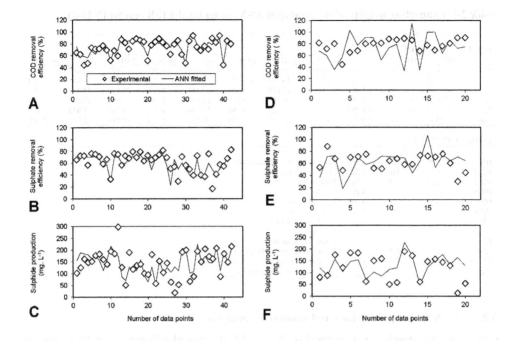

Figure 3-9. Experimental and ANN model fitted values

(A) COD removal, (B) sulfate removal and (C) sulfide production in the IFB for the training data (NT$_r$ = 42). Experimental and ANN model fitted values of (D) COD removal, (E) sulfate removal and (F) sulfide production in the IFB for the test data (NT$_e$ = 20)

Table 3-6. Sensitivity analysis of input variables for predicting COD and sulphate removal efficiency and sulphide production profiles in IFB R2

Input variable	Absolute average sensitivity (AAS) values		
	COD removal (%)	Sulphate removal (%)	Sulphide production (mg. L^{-1})
HRT (d)	0.058	0.137	0.132
COD (mg. L^{-1})	0.154	0.143	0.151
SO$_4^{2-}$ (mg. L^{-1})	0.374	0.182	0.310
Biomass (mg. L^{-1})	0.200	0.199	0.121
DpH	0.213	0.338	0.284

Table 3-7. Connection weights of the developed ANN model for the IFB reactor (5-11-3)

INPUT LAYER TO HIDDEN LAYER

	HID1	HID2	HID3	HID4	HID5	HID6	HID7	HID8	HID9	HID10	HID11
X_1	-5.841	-10.761	-27.416	-2.581	0.858	4.955	-2.662	-7.787	31.293	-1.648	13.015
X_2	-7.088	0.526	-7.890	-21.914	-16.299	-27.625	-6.271	-15.024	-15.297	-44.013	-17.530
X_3	-22.564	25.127	31.159	1.894	15.888	-4.047	-27.144	9.113	-21.619	35.487	-11.085
X_4	38.016	-15.727	35.691	10.868	-20.391	45.299	16.526	4.899	8.980	-27.567	-46.600
X_5	12.443	-15.005	-17.254	16.811	4.225	-8.786	26.286	6.726	-21.702	25.853	40.452
Bias	-0.794	-4.844	-10.741	-2.340	1.753	1.766	-0.232	-0.798	15.794	2.710	4.881

HIDDEN LAYER TO OUTPUT LAYER

	Y_1	Y_2	Y_3	
HID1	-3.590	-3.766	0.799	HID1 to HID11 - Hidden layer neurons
HID2	-5.937	-0.446	3.281	Bias - Bias term
HID3	1.725	0.238	-1.568	Input to the model
HID4	-4.669	-8.572	-5.166	X_1 - HRT (d)
HID5	5.537	1.324	0.605	X_2 - COD (mg. L^{-1})
HID6	0.139	4.882	1.764	X_3 - SO_4^{2-} (mg. L^{-1})
HID7	3.091	-2.819	-3.155	X_4 - Biomass (mg. L^{-1})
HID8	2.239	-1.055	1.423	X_5 - DpH
HID9	0.300	6.551	5.062	Output of the model
HID10	0.175	-1.128	-1.912	Y_1 - COD removal efficiency (%)
HID11	-15.833	5.296	1.374	Y_2 - Sulphate removal efficiency (%)
Bias	0.761	0.731	0.010	Y_3 - S^{2-} concentrations (mg. L^{-1})

3.3.5   ANN model predictions and sensitivity analysis

Concerning the training data predictions for the COD removal efficiency, sulphate removal efficiency and sulphide production, except for some outliers, the ANN model was able to map the behaviour of all the process outputs from the IFB (Figure 3-9A-C). It can be seen that the values of the COD removal efficiency < 55%, sulphate removal efficiency > 75% and sulphide production < 50 mg/L were not adequately mapped by the ANN model. During the testing phase (Figure 3-9D-F), the model was able to generalize the output variables and show an average of the performance trend of the reactor. Evidently, the model was not able to map the very low and very high peaks of the output variables during model testing. The ANN model with the configuration 5-11-3 was able to predict the COD and sulphate removal efficiencies and sulphide concentrations with R^2 values < 0.46.

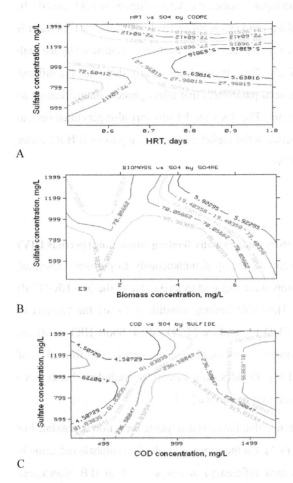

A

B

C

Figure 3-10. The effect of different input parameters on sulphate reduction
(A) COD removal efficiency, (B) sulphate removal efficiency and (C) sulphide production.

The AAS results for the developed model are shown in Table 3-6, while the different connection weights are given in

Table 3-7, *i.e.* the connection weights between the input to hidden layers and hidden to output layers. Using sulphate concentration as an input variable, the AAS values were 0.374 and 0.310 for the COD removal efficiency and sulphide production, respectively. Besides, DpH of the IFB also affected the sulphate removal efficiency and sulphide production as evidenced by the high ASS values of 0.338 and 0.284, respectively. Thus, these results clearly show that the influent sulphate concentration and the pH are crucial parameters that affect process intensification as well as the performance of an IFB bioreactor. The ANN model software also generated several contour plots to map the relationships between the model inputs and outputs of IFB R2 under transient feeding conditions (Figure 3-10A-C).

3.4 Discussion

3.4.1 Performance of the IFB bioreactors under steady feeding conditions (periods I-IV)

Both reactors, IFB R1 and IFB R2, were started up simultaneously to perform biological sulphate reduction. In period I, the reactors were exposed to stress due to the low HRT (1 d) and low influent COD concentrations. The COD limiting conditions forced the bacteria to perform sulphate reduction under low COD to sulphate (0.71) ratios. Short HRT (1 d) are usually preferred to outcompete planktonic microorganism and improve the formation of biofilms during bioreactor start up [30,31]. Besides, low density polyethylene beads have shown good capabilities for growing biofilm on its surface [16].

Several bioreactor studies have demonstrated that biological sulphate reduction is possible but not efficient at low COD:sulphate ratios (< 1). On the contrary, biological sulphate reduction is very efficient ($\geq$ 85% of sulphate removal efficiency) at ratios > 1.5 in IFB bioreactors [13,16,17]. The VSS concentration decreased from > 800 and < 1,000 mg VSS.L^{-1} to ~ 150 mg VSS.L^{-1} in the IFB R1 and IFB R2, respectively, from day 0 of operation until the end of period I. Due to a drop in the VSS concentration, the SRB population size as well as other bacteria and archaea present in the consortia could have also been simultaneously reduced. During period I, the sulphate removal efficiency was 20 ($\pm$ 10)% and 16 ($\pm$ 14)% for IFB R1 and IFB R2, respectively. The loss of SRB biomass, COD limitation and the prevailing hydrodynamic conditions in the IFB influenced sulphate removal during period I (Figure 3-5B and Figure 3-6B, period I).

The COD removal rates increased linearly from 341 ($\pm$ 101) to 1,410 ($\pm$ 228) mg COD.L^{-1}d^{-1}, as a function of the COD loading rate (530-2,000 mg COD.L^{-1}d^{-1}). These removal rates corresponded to IFB performances ranging from 65 to 86% for COD removal (Figure 3-7A and Figure 3-8A). Anew, the sulphate removal rates were nearly similar in both IFB bioreactors, ranging between 121 ($\pm$ 100) to 1,090 ($\pm$ 107) mg $SO_4^{2-}.L^{-1}d^{-1}$ during periods I-IV. The sulphate removal rates were not limited or inhibited by the sulphate loading rate (746-1,492 mg $SO_4^{2-}.L^{-1} d^{-1}$) and the sulphate removal efficiencies mostly averaged ~70% (Figure 3-7B and Figure 3-8B).

In an ideal process operation scenario, one would like to achieve $\geq$ 99% removal efficiency for COD and sulphate at a COD to sulphate ratio $\leq$ 1 during the IFB operation. These efficiencies might be possible with pure cultures of SRB, but not with the (syntrophic) microorganisms that grow in mixed microbial communities as present in anaerobic sludge. Bioreactors seeded with anaerobic sludge require more COD to remove sulphate at high efficiencies. For instance, COD to sulphate ratios of 2-2.5 were recommended to achieve sulphate removal efficiencies in the order of $\geq$ 90% [32,33]. From an application view point, it is always advisable to inoculate the reactor with mixed cultures rather than with pure cultures because mixed cultures can adapt to the varying wastewater composition without long start up times and are able to handle fluctuations in wastewater composition or loads.

Although the COD:sulphate ratios used in both IFB reactors during periods II, III and IV varied between 1.34 and 1.82, it did not majorly affect neither the COD nor the sulphate removal efficiencies. Furthermore, a closer look at the data on the sulphate removal rate (121 $\pm$ 100 to 1,090 $\pm$ 107 mg $SO_4^{2-}.L^{-1}d^{-1}$) and the COD loading rate (530-2,000 mg $COD.L^{-1}d^{-1}$) shows that the sulphate reduction process depended ~50% on the COD loading rate in both IFB bioreactors (Figure 3-7C and Figure 3-8C). Several studies have also reported the effects of COD limitation and the influence of the COD:sulphate ratio on the sulphate removal in IFB reactors [13,16,17]. However, it is noteworthy to mention that the rate and efficiency of biological sulphate reduction depends on the type of electron donor used, the pH, temperature, reactor configuration and the competition between SRB and other bacterial species present in the bioreactor [13,14,16,17,34].

3.4.2   Effect of transient feeding conditions on IFB bioreactor operation

IFB R2 was subjected to transient feeding conditions during period V, as feast (V-A, 4,265 mg $COD.L^{-1}d^{-1}$ and 2,990 mg $SO_4^{2-}.L^{-1}d^{-1}$) and famine (V-B, 0 mg $COD.L^{-1}d^{-1}$ and 0 mg $SO_4^{2-}.L^{-1}d^{-1}$) conditions. Under such feeding conditions, the bacterial consortia experience physiological stress and undergo complex cellular level interactions between starvation mediated metabolic process and stress resistance responses. Usually, this is an operational challenge for full scale wastewater treatment systems. In this study, IFB R2 was expected to fail biologically in terms of COD as well as sulphate removal performance after a few days of transient operation. However, interestingly, the reactor was able to withstand over 10 successive transient feeding conditions for 66 d. The COD (445 ± 93 mg.L^{-1}), sulphate (510 ± 238 mg.L^{-1}) and sulphide (300 ± 49 mg.L^{-1}) concentration in the effluent of IFB R2 reached nearly twice the concentration during the feast period (V-A) when compared to the previous period (IV) of operation of IFB R2 (Figure 3-6) and to the control reactor IFB R1 during period V (Figure 3-5). The opposite was observed during the famine period, wherein the COD, sulphate and sulphide concentrations in the effluent were negligible (Figure 3-6, period V-B). Despite being subjected to alternate cycles of feast or famine conditions, the VSS profiles in the IFB were maintained at ~260 mg.L^{-1}.

In addition to the resilience capacity and tolerance of the SRB to feast and famine conditions, the fast response of the IFB and its ability to maintain good COD and sulphate removal efficiencies can be attributed to the mixing characteristics of the IFB bioreactors [35]. The mixing capacity was determined with the help of RTD analysis (Figure 3-4) and the results showed the same characteristics as that of a CSTR. The tracer took almost 4 times the HRT to completely leave the IFB bioreactor, and under such conditions, the reactors usually behave as a very large CSTR [36]. The RTD analysis demonstrated that after 3.9 times the HRT, $i.e.$ $3.9 \times \theta$, and during the famine period of IFB R2 operation, the concentrations of the electron donor (COD-lactate), electron acceptor (sulphate) and end product (sulphide) of this microbial mediated reaction reaches negligible values.

IFB R1 performed pseudo steady state sulphate reduction ($61 \pm 15\%$) during period V (Figure 3-5B), even under continuous and steady feeding conditions, $i.e.$ 2,000 mg COD.L^{-1}d^{-1} and 1,492 mg SO$_4^{2-}$.L^{-1}d^{-1}. Presumably, this behaviour might be due to the long term competition of the SRB against the hydrolytic-fermentative bacteria for lactate [37,38], rather than the competition of SRB and methanogens for acetate. Acetate was not fed and therefore it was not the first available electron donor for SRB. The low affinity of SRB for acetate has been demonstrated in the literature by several researchers [18,39]. For instance, Janyasuthiwong et $al.$ [18] demonstrated in batch bioreactors that the sulphate removal efficiency decreased from ~ 28% to 8% with an increase in the acetate concentration from 100 to 500 mg.L^{-1}. O'Reilly and Colleran [39] conducted activity tests with anaerobic sludge, using acetate and sulphate as the substrates in the presence of a selective inhibitor for methanogenic archaea. Their results indicated negligible acetate consumption ($\sim$ 100 mg.L^{-1}) and negligible methane production ($\geq$ 2 mL biogas), irrespective of the COD:sulphate ratio tested (16, 4 and 2), hence confirming the lack of acetate degrading SRB.

3.4.3 Robustness of biological sulphate reduction in IFB bioreactors

Robustness is the capacity of a system to overcome an unexpected perturbation without failing and its ability to continue to demonstrate steady state performance [40]. In this study, the robustness of sulphate reduction in an IFB bioreactor was demonstrated in IFB R2: the bacteria in IFB R2 were capable to overcome the presence of excess substrate (feast) and survive in the absence of the substrate (famine). The average values of COD (3,330 $\pm$ 963 mg COD.L^{-1}d^{-1}) and sulphate (2,095 $\pm$ 519 mg SO$_4{}^{2-}$.L^{-1}d^{-1}) removal rates under feast conditions (period V-A) remained at the same performance level as in the previous period and as in the control reactor IFB R1 (COD and sulphate removal: $\sim$ 70%). Nevertheless, as indicated earlier, during the famine period, negative values were achieved, *i.e.* -70 ($\pm$ 282)% and -552 ($\pm$ 928)% for the COD and sulphate removal efficiency (Table 3-2), respectively. Such negative removal values are not uncommon in bioreactors and this can be mainly attributed to the influence of reactor hydrodynamics (3.9$\times\theta$) and the wash out of electron donor and/or acceptor that were not used by the bacteria.

The profile of sulphide during the famine periods (Figure 3-6C, period V-B) show the same trend of dilution as the RTD profile (Figure 3-4) and the sulphide was diluted in the IFB R2 effluent in a time equivalent to 3.9$\times\theta$. Furthermore, during the famine period, the sulphide production was less (68 $\pm$ 65 mg S^{2-}.L^{-1}) when compared to the feast period (300 $\pm$ 49 mg S^{2-}.L^{-1}), Figure 3-8D. This confirms that little residual COD and sulphate present were used to produce sulphide in the IFB R2 during the famine phase, according to the following stoichiometric equation [41]:

$$2\text{Lactate}^- + SO_4{}^{2-} \rightarrow 2\text{Acetate}^- + HS^- + 2HCO_3{}^- + H^+ \qquad \text{Eq. 3-16}$$

Given this observation, the hypothesis of production of sulphide from carbon and sulphur accumulated intracellularly in the form of polythioester or polyhydroxyalkanoate is discarded. SRB do not produce polyhydroxyalkanoates when lactate is fed as substrate [42]. This contrasts the findings from another research with IFB bioreactors, but with a slight difference in its construction (the recirculation was installed in the middle of the IFB bioreactor) and therefore different RTD profiles and mixing properties [43]. Villa-Gomez *et al.* [43] reported an increasing production of sulphide during the famine period (the organic loading rate was decreased by half with respect to the feast period), and therefore hypothesized that COD was accumulated as storage products during the feast stage and later consumed during the famine stage.

The ability of the SRB to overcome successive famine periods and to restore the original performance during the feast periods clearly demonstrates that the biological sulphate reduction was 100% resilient in IFB R2. This is also evident from the good correlation between the loading rates and the respective volumetric removal rates, where the R^2 values were > 0.95 for IFB R2 (Figure 3-8A-C).

3.4.4 ANN modelling and transient feeding conditions

Figure 3-10A shows that if the HRT is < 0.65 d and the influent sulphate concentration ranges between 600-850 mg.L^{-1}, the COD removal efficiencies will be > 72%. However, according to the model predictions, to achieve > 94% COD removal, the HRT should be further decreased. This prediction supports the information obtained by the RTD analysis on the adaptation time required and the relation of the specific growth rate (μ, d^{-1}) with the HRT (Eq. 3-17):

$$\mu = 1/HRT \hspace{5cm} \text{Eq. 3-17}$$

Hence, microorganisms with fast growth rates prevail inside the reactor at short HRT rather than those which have low growth rates. Therefore, this hydrodynamic stress could influence the development of a robust biofilm that hosts microorganisms capable of carrying out different enzymatic biochemical reactions [30,44,45]. On the other hand, HRT values > 0.65 d and influent sulphate concentrations > 1,300 mg.L^{-1} also promoted high COD removal efficiencies (> 94%) in the IFB. Such high sulphate concentrations in the influent can increase the flow of carbon during the biological sulphate reduction and also facilitates the selection of more efficient biochemical pathways for the utilization of carbon [46].

The sensitivity analysis (Table 3-6) showed the relationships between the COD removal and influent sulphate concentration (AAS = 0.374), the sulphate removal and DpH (AAS = 0.338) and the sulphide production and influent sulphate concentration (0.310). The AAS values show that biomass concentration also played a complementary role in determining the performance of the IFB. Additionally, according to Figure 3-10B, varying biomass concentrations of 1,000 to 6,000 mg.L^{-1} in the reactor and influent sulphate concentrations < 1,000 mg.L^{-1} will not hamper the sulphate removal and the IFB might perform at efficiencies > 90%. Such conditions of substrate (influent sulphate) and biomass concentration will yield a substrate:biomass ratio of ~1.0 to 0.16. The substrate:biomass ratio can influence the substrate removal rate by following a zero or first order behaviour. From a practical perspective, biomass concentration, mode of biomass growth in the bioreactor (attached or suspended) and microbial activity are critical parameters which determine the reactor performance for wastewater treatment [47].

The ANN model also predicted the COD (~ 600 - 1600 mg.L^{-1}) and sulphate (~ 400 - 1000 mg.L^{-1}) concentrations that have to be fed to the IFB to produce 313 mg.L^{-1} of sulphide (Figure 3-10C). This was also supported by the results of the sensitivity analysis: the sulphate concentration was a strong variable affecting the COD removal (0.374) and the sulphide production (0.310). The predicted sulphide concentration (313 mg.L^{-1}) by the ANN was very similar to the sulphide concentrations produced during the feast period (300 ± 49 mg.L^{-1}, Figure 3-8D). The predicted sulphide concentration represented ~940 mg.L^{-1} of sulphate reduced and a 63% sulphate removal efficiency. This was close to the sulphate removal efficiency observed in IFB R2 during the feast period (67 ± 15%, Table 3-2, period V-A) and also comparable to the sulphate removal efficiency in the control reactor, IFB R1 (61 ± 15%, Table 3-2, period V). Sulphate removal usually depends on the COD:sulphate ratio. However, high residual COD in the treated water is not recommended. Several literature reports have suggested performing biological sulphate reduction at a COD:sulphate ratio close to 0.67 in order to reduce operational costs [13,16,17].

As shown previously in Eq. 3-16, the reduction of sulphate and partial oxidation of COD (lactate) results in the formation of acetate, sulphide, carbonate and protons. However, many other reactions can also occur simultaneously with the intermediates formed during the utilization of COD (lactate) by bacteria other than SRB [41]. Lactate can be used by hydrolytic-fermentative bacteria to produce propionate or ethanol [34] and subsequently these compounds were used by SRB as electron donor for sulphate reduction to produce acetate, carbonate and protons [41]. In such a food chain, the final by-products like acetate, carbonate and protons are substrates that are most likely used by methanogenic archaea and could also play an important role in maintaining the buffering capacity of the IFB bioreactors.

The influent pH was ~ 5.5-6.1, while the pH in the IFB bioreactor was ~ 7.1-7.4 (Figure 3-7E and Figure 3-8E). The results of the sensitivity analysis showed that DpH strongly affected the sulphate removal (0.338) followed by sulphide production (0.284). This clearly suggests that the sulphate reduction process was mainly responsible for the changes of pH and buffer capacity of the IFB bioreactors.

3.5 Conclusions

Two IFB bioreactors were started simultaneously to perform sulphate reduction under continuous operation (IFB R1, control reactor) and transient feeding (IFB R2, feast and famine) conditions. Sulphate removal in IFB R2 was robust and resilient to transient feeding conditions. The removal efficiency of sulphate during the feast period ($67 \pm 15\%$) was similar to that of the same IFB R2 under steady feeding conditions ($71 \pm 4\%$) and the control reactor IFB R1 ($61 \pm 15\%$). A three-layered ANN model (5-11-3) was successfully developed and tested to forecast the performance parameters of the IFB, namely the COD removal efficiency, sulphate removal efficiency and sulphide production. Results from the sensitivity analysis showed that the influent sulphate concentrations affected both the COD removal efficiency and the sulphide production, also pH changes (from acid to neutral) were induced by the sulphate removal process during the IFB operation.

3.6 References

[1]   A.M. Sarmiento, M. Olías, J.M. Nieto, C.R. Cánovas, J. Delgado, Natural attenuation processes in two water reservoirs receiving acid mine drainage, Sci. Total Environ. 407 (2009) 2051–2062. doi:10.1016/j.scitotenv.2008.11.011.

[2]   H. Bai, Y. Kang, H. Quan, Y. Han, J. Sun, Y. Feng, Treatment of acid mine drainage by sulfate reducing bacteria with iron in bench scale runs, Bioresour. Technol. 128 (2013) 818–822. doi:10.1016/j.biortech.2012.10.070.

[3]   C. Monterroso, F. Macías, Drainage waters affected by pyrite oxidation in a coal mine in Galicia (NW Spain): Composition and mineral stability, Sci. Total Environ. 216 (1998) 121–132. doi:10.1016/S0048-9697(98)00149-1.

[4]   F. Mapanda, G. Nyamadzawo, J. Nyamangara, M. Wuta, Effects of discharging acid-mine drainage into evaporation ponds lined with clay on chemical quality of the surrounding soil and water, Phys. Chem. Earth, Parts A/B/C. 32 (2007) 1366–1375. doi:10.1016/j.pce.2007.07.041.

[5]   R.M.M. Sampaio, R.A. Timmers, Y. Xu, K.J. Keesman, P.N.L. Lens, Selective precipitation of Cu from Zn in a pS controlled continuously stirred tank reactor, J. Hazard. Mater. 165 (2009) 256–265. doi:10.1016/j.jhazmat.2008.09.117.

[6]   J. Dries, A. De Smul, L. Goethals, H. Grootaerd, W. Verstraete, High rate biological treatment of sulfate-rich wastewater in an acetate-fed EGSB reactor, Biodegradation. 9

(1998) 103–111. doi:10.1023/A:1008334219332.

[7]   P. Kijjanapanich, A.T. Do, A.P. Annachhatre, G. Esposito, D.H. Yeh, P.N.L. Lens, Biological sulfate removal from construction and demolition debris leachate: Effect of bioreactor configuration, J. Hazard. Mater. 269 (2014) 38–44. doi:10.1016/j.jhazmat.2013.10.015.

[8]   D. Deng, L.-S. Lin, Two-stage combined treatment of acid mine drainage and municipal wastewater, Water Sci. Technol. 67 (2013) 1000–1007. doi:10.2166/wst.2013.653.

[9]   F.J. Torner-Morales, G. Buitron, Kinetic characterization and modeling simplification of an anaerobic sulfate reducing batch process, J. Chem. Technol. Biotechnol. 85 (2010) 453–459. doi:10.1002/jctb.2310.

[10]   M.V.G. Vallero, J. Sipma, G. Lettinga, P.N.L. Lens, High-rate sulfate reduction at high salinity (up to 90 mS.cm^{-1}) in mesophilic UASB reactors., Biotechnol. Bioeng. 86 (2004) 226–35. doi:10.1002/bit.20040.

[11]   O.B.D. Thabet, H. Bouallagui, J. Cayol, B. Ollivier, M.-L. Fardeau, M. Hamdi, Anaerobic degradation of landfill leachate using an upflow anaerobic fixed-bed reactor with microbial sulfate reduction, J. Hazard. Mater. 167 (2009) 1133–1140. doi:10.1016/j.jhazmat.2009.01.114.

[12]   J. Sipma, M.B. Osuna, G. Lettinga, A.J.M. Stams, P.N.L. Lens, Effect of hydraulic retention time on sulfate reduction in a carbon monoxide fed thermophilic gas lift reactor., Water Res. 41 (2007) 1995–2003. doi:10.1016/j.watres.2007.01.030.

[13]   L.B. Celis-García, E. Razo-Flores, O. Monroy, Performance of a down-flow fluidized bed reactor under sulfate reduction conditions using volatile fatty acids as electron donors, Biotechnol. Bioeng. 97 (2007) 771–779. doi:10.1002/bit.21288.

[14]   L.B. Celis, D. Villa-Gómez, A.G. Alpuche-Solís, B.O. Ortega-Morales, E. Razo-Flores, Characterization of sulfate-reducing bacteria dominated surface communities during start-up of a down-flow fluidized bed reactor., J. Ind. Microbiol. Biotechnol. 36 (2009) 111–21. doi:10.1007/s10295-008-0478-7.

[15]   M. Gallegos-Garcia, L.B. Celis, R. Rangel-Méndez, E. Razo-Flores, Precipitation and recovery of metal sulfides from metal containing acidic wastewater in a sulfidogenic down-flow fluidized bed reactor, Biotechnol. Bioeng. 102 (2009) 91–99. doi:10.1002/bit.22049.

[16]   D. Villa-Gomez, H. Ababneh, S. Papirio, D.P.L. Rousseau, P.N.L. Lens, Effect of sulfide concentration on the location of the metal precipitates in inversed fluidized bed reactors, J. Hazard. Mater. 192 (2011) 200–207. doi:10.1016/j.jhazmat.2011.05.002.

[17]   S. Papirio, G. Esposito, F. Pirozzi, Biological inverse fluidized-bed reactors for the treatment of low pH- and sulphate-containing wastewaters under different COD/SO_4^{2-} conditions., Environ. Technol. 34 (2013) 1141–9. doi:10.1080/09593330.2012.737864.

[18]   S. Janyasuthiwong, E.R. Rene, G. Esposito, P.N.L. Lens, Effect of pH on the performance of sulfate and thiosulfate-fed sulfate reducing inverse fluidized bed reactors, J. Environ. Eng. 142 (2015) C4015012. doi:10.1061/(ASCE)EE.1943-7870.0001004.

[19]   S. Bardestani, M. Givehchi, E. Younesi, S. Sajjadi, S. Shamshirband, D. Petkovic, Predicting turbulent flow friction coefficient using ANFIS technique, Signal, Image Video Process. (2016). doi:10.1007/s11760-016-0948-8.

[20]   A. Gani, A. Siddiqa, S. Shamshirband, F. Hanum, A survey on indexing techniques for big data: taxonomy and performance evaluation, Knowl. Inf. Syst. 46 (2016) 241–284. doi:10.1007/s10115-015-0830-y.

[21]   L. Olatomiwa, S. Mekhilef, S. Shamshirband, D. Petković, Adaptive neuro-fuzzy approach for solar radiation prediction in Nigeria, Renew. Sustain. Energy Rev. 51 (2015) 1784–1791. doi:10.1016/j.rser.2015.05.068.

[22]   S. Sajjadi, S. Shamshirband, M. Alizamir, P.L. Yee, Z. Mansor, A.A. Manaf, T.A. Altameem, A. Mostafaeipour, Extreme learning machine for prediction of heat load in district heating systems, Energy Build. 122 (2016) 222–227. doi:10.1016/j.enbuild.2016.04.021.

[23]   APHA, Standard Methods for the Examination of Water and Wastewater, 20th ed., Washington, USA., 1999.

[24]   C. Sheli, R. Moletta, Anaerobic treatment of vinasses by a sequentially mixed moving bed biofilm reactor, Water Sci. Technol. 56 (2007) 1–7. doi:10.2166/wst.2007.465.

[25]   H. Scott Fogler, Elements of chemical reaction engineering, 3rd ed., Upper Saddle River, N.J. : Prentice Hall PTR, 1987. doi:10.1016/0009-2509(87)80130-6.

[26]   E.R. Rene, M.E. López, M.C. Veiga, C. Kennes, Artificial neural network modelling for waste, in: B. Igelnik (Ed.), Comput. Model. Simul. Intellect Curr. State Futur. Perspect.,

IGI Global, Hershey PA, USA, 2011: pp. 224–263. doi:10.4018/978-1-60960-551-3.ch010.

[27]   H.R. Maier, G.C. Dandy, The effect of internal parameters and geometry on the performance of back-propagation neural networks: An empirical study, Environ. Model. Softw. 13 (1998) 193–209. doi:10.1016/S1364-8152(98)00020-6.

[28]   D.E. Rumelhart, G.E. Hinton, R.J. Williams, Parallel distributed processing: explorations in the microstructure of cognition, Vol. 1, in: D.E. Rumelhart, J.L. McClelland, C. PDP Research Group (Eds.), MIT Press, Cambridge, MA, USA, 1986: pp. 318–362.

[29]   E.R. Rene, M.E. López, H.S. Park, D.V.S. Murthy, T. Swaminathan, ANNs for identifying shock loads in continuously operated biofilters, in: Handb. Res. Ind. Informatics Manuf. Intell., IGI Global, 2012: pp. 72–103. doi:10.4018/978-1-4666-0294-6.ch004.

[30]   R. Cresson, R. Escudié, J.P. Steyer, J.P. Delgenès, N. Bernet, Competition between planktonic and fixed microorganisms during the start-up of methanogenic biofilm reactors, Water Res. 42 (2008) 792–800. doi:10.1016/j.watres.2007.08.013.

[31]   R. Escudié, R. Cresson, J.-P. Delgenès, N. Bernet, Control of start-up and operation of anaerobic biofilm reactors: An overview of 15 years of research, Water Res. 45 (2011) 1–10. doi:10.1016/j.watres.2010.07.081.

[32]   F.J. Torner-Morales, G. Buitrón, Kinetic characterization and modeling simplification of an anaerobic sulfate reducing batch process, J. Chem. Technol. Biotechnol. 85 (2009) 453–459. doi:10.1002/jctb.2310.

[33]   A. Velasco, M. Ramírez, T. Volke-Sepúlveda, A. González-Sánchez, S. Revah, Evaluation of feed COD/sulfate ratio as a control criterion for the biological hydrogen sulfide production and lead precipitation, J. Hazard. Mater. 151 (2008) 407–413. doi:10.1016/j.jhazmat.2007.06.004.

[34]   W. Liamleam, A.P. Annachhatre, Electron donors for biological sulfate reduction, Biotechnol. Adv. 25 (2007) 452–463. doi:10.1016/j.biotechadv.2007.05.002.

[35]   T. Renganathan, K. Krishnaiah, Liquid phase mixing in 2-phase liquid–solid inverse fluidized bed, Chem. Eng. J. 98 (2004) 213–218. doi:10.1016/j.cej.2003.08.001.

[36]   A. Martin, Interpretation of residence time distribution data, Chem. Eng. Sci. 55 (2000)

5907–5917. doi:10.1016/S0009-2509(00)00108-1.

[37]   A. Wang, N. Ren, X. Wang, D. Lee, Enhanced sulfate reduction with acidogenic sulfate-reducing bacteria, J. Hazard. Mater. 154 (2008) 1060–1065. doi:10.1016/j.jhazmat.2007.11.022.

[38]   S.A. Dar, R. Kleerebezem, A.J.M. Stams, J.G. Kuenen, G. Muyzer, Competition and coexistence of sulfate-reducing bacteria, acetogens and methanogens in a lab-scale anaerobic bioreactor as affected by changing substrate to sulfate ratio, Appl. Microbiol. Biotechnol. 78 (2008) 1045–1055. doi:10.1007/s00253-008-1391-8.

[39]   C. O'Reilly, E. Colleran, Effect of influent COD/SO$_4^{2-}$ ratios on mesophilic anaerobic reactor biomass populations: physico-chemical and microbiological properties, FEMS Microbiol. Ecol. 56 (2006) 141–153. doi:10.1111/j.1574-6941.2006.00066.x.

[40]   H. Kitano, Towards a theory of biological robustness, Mol. Syst. Biol. 3 (2007) 1–7. doi:10.1038/msb4100179.

[41]   Y. Zhao, N. Ren, A. Wang, Contributions of fermentative acidogenic bacteria and sulfate-reducing bacteria to lactate degradation and sulfate reduction, Chemosphere. 72 (2008) 233–242. doi:10.1016/j.chemosphere.2008.01.046.

[42]   T. Hai, D. Lange, R. Rabus, A. Steinbuchel, Polyhydroxyalkanoate (PHA) accumulation in sulfate-reducing bacteria and identification of a class III *PHA synthase* (PhaEC) in *Desulfococcus multivorans*, Appl. Environ. Microbiol. 70 (2004) 4440–4448. doi:10.1128/AEM.70.8.4440-4448.2004.

[43]   D.K. Villa-Gomez, J. Cassidy, K.J. Keesman, R. Sampaio, P.N.L. Lens, Sulfide response analysis for sulfide control using a pS electrode in sulfate reducing bioreactors, Water Res. 50 (2014) 48–58. doi:10.1016/j.watres.2013.10.006.

[44]   P. Verhagen, L. De Gelder, S. Hoefman, P. De Vos, N. Boon, Planktonic versus biofilm catabolic communities: Importance of the biofilm for species selection and pesticide degradation, Appl. Environ. Microbiol. 77 (2011) 4728–4735. doi:10.1128/AEM.05188-11.

[45]   N. Barkai, S. Leibler, Robustness in simple biochemical networks., Nature. 387 (1997) 913–7. doi:10.1038/43199.

[46]   K.S. Habicht, L. Salling, B. Thamdrup, D.E. Canfield, Effect of low sulfate concentrations on lactate oxidation and isotope fractionation during sulfate reduction by

Archaeoglobus fulgidus strain Z., Appl. Environ. Microbiol. 71 (2005) 3770–7. doi:10.1128/AEM.71.7.3770-3777.2005.

[47]   Y. Liu, The S_0/X_0-dependent dissolved organic carbon distribution in substrate-sufficient batch culture of activated sludge, Water Res. 34 (2000) 1645–1651. doi:10.1016/S0043-1354(99)00293-6.

Chapter 4

High rate biological sulphate reduction in a lactate fed inverse fluidized bed reactor at a hydraulic retention time of 3 h

A modified version of this chapter has been submitted to Bioprocess and Biosystems Engineering journal as:
Reyes-Alvarado L.C., Hatzikioseyian A., Rene E.R., Houbron E., Rustrian E., Esposito G., and Lens P.N.L., (2017). Hydrodynamic studies and model development for biological sulphate reduction using lactate as electron donor in an inverse fluidized bed bioreactor at a hydraulic retention time of up to 3 h

Abstract

Industrial wastewater rich in sulphate might cause acidification of the receiving water bodies and have toxic, corrosive and malodorous effects when sulphide is produced. In this research, an inverse fluidized bed bioreactor (IFBB) was operated at a decreasing hydraulic retention time (HRT) of 1-0.125 d for 155 d divided in 8 periods. The characteristics of the influent were: sulphate concentration 745 ($\pm$ 17) mg.L^{-1}, COD:SO$_4^{2-}$ ratio of 1.2-2.4, COD was supplied as lactate and pH 5.2-6.2. The highest removal rates were 2,646 and 4,866 mg SO$_4^{2-}$.L^{-1}d^{-1} using a COD:SO$_4^{2-}$ ratio of 2.3 at an HRT of 0.25 and 0.125 d, respectively. The biological sulphate reduction was limited by the influent COD concentrations at a COD:SO$_4^{2-}$ ratio < 2.3. The IFBB behaved hydrodynamically as a continuous stirred tank reactor (CSTR) and ensured biomass retention at a maximum residence time of θ = 3.84 ($\pm$ 0.013), according to the RTD analysis. The Grau second order substrate removal model described the biological sulphate reduction (R^2 > 0.96) under the conditions tested. The IFBB, with a sulphate removal efficiency > 75% at an HRT < 0.25 d, is thus a promising reactor configuration for practical purposes.

Keywords: inverse fluidized bed bioreactor (IFBB), sulphate reduction, industrial wastewater, high rate removal

4.1 Introduction

Industrial wastewaters rich in sulphate (≈ 0.4-21 g.L^{-1}), for instance from the mining and metallurgical industry, contains high concentrations of heavy metals, very low pH, high redox potential, but have a very low chemical oxygen demand (COD) concentration [1,2]. Acid mine drainage can damage the flora and fauna of water reservoirs [3], groundwater and land [4]. At environmental conditions, sulphate can be reduced by sulphate reducing bacteria (SRB) to sulphide, which is a corrosive, hazardous and toxic weak acid that is volatile at low pH and induces serious problems to human and animal beings already at low concentrations [5].

Wastewater treatment processes require sustainable solutions at low construction and operation costs. In sulphate rich industrial wastewaters, the lack of COD is a disadvantage for the biological treatment that requires the addition of expensive electron donors for the reduction of sulphate. Recently, the trend of coupling multiple processes in compact bioreactors is developing as a process intensification (PI) strategy [6], e.g. the use of inverse fluidized bed bioreactors (IFBB) for coupling sulphate reduction and metal-sulphide precipitation and recovery of the metals from the treated acid mine drainage in a single reactor [7]; or the sulphate reduction process coupled to the decolourization of azo dyes in sequencing batch bioreactors [8]. The size (volume) of the bioreactor can be reduced if high rate bioprocesses are developed, thus reducing the costs of construction and operation.

Two and three phase contact IFBB have been extensively studied, i.e. liquid-solid [9] and liquid-solid-gas [10,11] respectively. This reactor type has been used for aerobic processes in COD reduction [12] and widely characterized for anaerobic digestion, more specific for methane production from wine distillery [13] and brewery [14] wastewater, as well as the simultaneous removal of nutrients as nitrogen and carbon from wastewater [15] or sulphate from construction demolition debris leachate [16]. Moreover, the IFBB has been studied under transient feeding condition under sulphidogenic conditions [17]. However, the biological sulphate reduction kinetics have not been evaluated in IFBB treating inorganic sulphate rich wastewater. Monod equations [18] are, most frequently, used to calculate the kinetic parameters for continuous biological sulphate removal processes [19,20]. But it is unknown if the sulphate reduction follows first or second order kinetics in the IFBB under any operational parameter or if it is affected by them. For instance, the first order substrate removal model fits the COD removal in a continuous up-flow anaerobic sludge blanket bioreactor ($R^2 = 0.93$) [21] and in an anaerobic hybrid bioreactor ($R^2 = 0.89$) treating fermentation based pharmaceutical wastewater [22]. Also, the Grau second order [23] and the Stover-Kincannon [24] substrate removal models are often used to evaluate the COD removal in anaerobic processes [21,22] but have not been used for applications as in IFBB and biological sulphate removal. COD removal was well described ($R^2 \geq 0.99$) with the use of Grau second order and Stover-Kincannon substrate removal models in a continuous up-flow anaerobic sludge blanket bioreactor [21] and in an anaerobic hybrid bioreactor treating fermentation based pharmaceutical wastewater [22].

To our knowledge, high rate sulphate reduction using high rate feeding conditions (HRT = 0.125 d) in an IFBB has not yet been reported. Therefore, it is necessary to evaluate the hydrodynamics (RTD analysis, bed expansion) of the IFBB and determine its sulphate reducing kinetics under high rate feeding conditions (HRT < 0.25).

4.2 Material and methods

4.2.1 Synthetic wastewater

Inorganic synthetic wastewater was used for the experiments in the IFB (in $mg.L^{-1}$): NH_4Cl (300), $MgCl_2 \cdot 6H_2O$ (120), KH_2PO_4 (200), KCl (250), $CaCl_2 \cdot 2H_2O$ (15), yeast extract (20) and 0.5 mL of a mixture of micronutrients [7]. The composition of trace elements was prepared with $FeCl_2 \cdot 4H_2O$ (1,500), $MnCl_2 \cdot 4H_2O$ (100), EDTA (500), H_3BO_3 (62), $ZnCl_2$ (70), $NaMoO_4 \cdot 2H_2O$ (36), $AlCl_3 \cdot 6H_2O$ (40), $NiCl_3 \cdot 6H_2O$ (24), $CoCl_2 \cdot 6H_2O$ (70), $CuCl_2 \cdot 2H_2O$ (20) and HCl 36 % (1 $mL.L^{-1}$). Sodium lactate was used as electron donor (COD) and sodium sulphate as electron acceptor. All reagents were of analytical grade. The pH was never controlled at neutral in the influent tank and, prior fed to the IFBB, it ranged from: 5.2 ($\pm$ 0.2) to 6.2 ($\pm$ 0.2) along the operation.

4.2.2 Inoculum

The inoculum was obtained from the anaerobic reactor digesting waste activated sludge at the municipal wastewater treatment plant at Harnaschpolder (The Netherlands) and contained 23.1 $g.L^{-1}$ of total suspended solids (TSS) and 16.1 $g.L^{-1}$ of volatile suspended solids (VSS). The IFBB was inoculated at 10 % of its active volume.

4.2.3 Carrier material

The carrier material was low density polyethylene beads (0.918 g. mL^{-1} at 25 °C, from Sigma Aldrich) with a diameter of 4 mm. Before the beads were used, they were rinsed with demineralised water in order to remove smaller fractions. To start-up the IFBB, 300 mL of the beads were placed inside the bioreactor and mixed with the inoculum.

4.2.4 Anaerobic inverse fluidized bed bioreactor

The IFBB reactor (effective volume of 2.46 L) was built with a transparent pipe manufactured with polyvinyl chloride (PVC, internal diameter 5.6 cm and length 103 cm) and consisted of a height to diameter ratio (*H/D*) of 17.8 with the same characteristics of that described by Reyes-Alvardo *et al.* [17]. The influent was supplied to the IFBB with a peristaltic pump (Masterflex L/S). The influent line was connected to the recirculation line that employed a recirculation pump (Iwaki Magnet Pump, Iwaki CO., LTD. Tokio, Japan). The IFBB mixed liquor was recirculated downwards at 122 m.h^{-1} down flow liquid velocity (DFLV) generated by 300 L.h^{-1} of recirculation flow velocity (RFV), measured with a flow meter. The outlet of the effluent was placed at 20 cm height from the bottom of the reactor. The liquid level inside the IFBB was kept constant by a simple level controller: the excess reactor liquid was displaced to the outer side by the feed. Neither a membrane nor a mesh was used at the bottom to keep the carrier inside the column and/or to prevent the suction by the recirculation pump.

4.2.5 Hydrodynamic evaluation of the IFB

4.2.5.1 Residence time distribution

The RTD was evaluated following the delta Dirac method as described in the literature [25]. This experiment was done by injecting a spike (2 mL) of a concentrated (1 M) sodium chloride solution. MiliQ water was used as eluent, the conductivity of this MiliQ water was used as base line. A calibration curve was made by diluting the concentrated salt solution and the conductivity of each dilution was the response to each concentration. The effluent conductivity was measured and converted into concentration. This analysis was made before the startup of the reactors and with a constant influent flow rate (Q_{in}) of 20 L. d^{-1} (HRT=0.125 d) and different recirculation flow velocities (200, 300 and 360 L.h^{-1}).

The function of the RTD (Eq. 4-1), *E(t)* with units of h^{-1}, was the quotient of the concentrations "*C(t)*" and the area below the curve of the profile of *C(t)* against time (Eq. 4-2). The integral of this denominator and further integrations were executed with the trapezoidal method using Microsoft Excel. The cumulative profile *F(t)* was obtained by the addition of each *E(t)* data evaluated at different times and corresponded to Eq. 4-3. The mean residence time (*t_m*) was evaluated by Eq. 4-4. The information concerning the distribution of the residence time was normalized to *E(θ)* using Eq. 4-5. Likewise, the time (*t*) was also normalized (*θ*) and defined by Eq. 4-6. Then, the profiles *E(θ)* against *θ* and *F(t)* against *θ* were constructed. More information concerning the procedure and interpretation of the RTD can be found elsewhere [25].

$$E(t) = C(t) / \int_0^\infty C(t)\, dt \qquad\qquad \text{Eq. 4-1}$$

$$\int_0^\infty C(t)\, dt \qquad\qquad \text{Eq. 4-2}$$

$$\int_0^t E(t)\, dt = F(t) \qquad\qquad \text{Eq. 4-3}$$

$$\int_0^\infty t E(t)\, dt = t_m \qquad\qquad \text{Eq. 4-4}$$

$$E(\theta) = E(t) \times t_m \qquad\qquad \text{Eq. 4-5}$$

$$\theta = t / t_m \qquad\qquad \text{Eq. 4-6}$$

4.2.5.2 Bed expansion

The bed expansion was measured along the column with different initial volumes of polyethylene beads (50, 100, 150, 200, 300, 350, 400, 450 mL) in MiliQ water. The static bed (H_0) was measured when no recirculation flow was applied. Each bed volume was tested at different recirculation (0, 200, 300, 350, 400 and 500 L.h^{-1}) or liquid down flow (0, 81, 122, 146, 162 and 203 m.h^{-1}) velocities inducing the expansion of the bed to a certain column high (H_f). The static and expanded bed were measured with a scale in centimeters and placed alongside the reactor.

The dimensionless number of the relative bed expansion (RBE) was evaluated as the ratio of the height of the static bed (H_0) or expanded bed (H_f) to the column height (H_C) as in Eq. 4-7. Since the polyethylene has buoyant properties due to its low density (0.918 g. mL^{-1} at 25 °C) compared to water (0.997 g. mL^{-1} at 25 °C), the top of the liquid inside the reactor was determined as zero and the bottom of the column of water was at 100 cm. The different results of the RBE were plotted against the DFLV of the liquid (m. h^{-1}) in the bioreactor column.

$$RBE = H_{0-f} / H_C \qquad\qquad \text{Eq. 4-7}$$

4.2.6 Reactor operation conditions

The IFB under sulphate reducing conditions was tested for 155 d and eight different periods of operation as described in Table 4-1. The reactor was started up in continuous mode from $t = 0$. The sulphate was fed at a concentration of 745 ($\pm$ 17) mg. L^{-1} and the COD as lactate was varied from 912 to 1,757 mg.L^{-1}. These concentrations of electron donor (lactate) and acceptor (sulphate) gave different COD:sulphate ratios (1.2-2.4). The influent pH was never controlled and ranged between 5.2-6.2. The influent tank was changed every 2-3 d. The liquid influent flow (Q_{in}) was increased (from 2.5 L. d^{-1} to 20 L.d^{-1}) in order to decrease in a stepwise mode the initial HRT from 1 d to 0.5, 0.25 and 0.125 d.

The sulphate reduction performance in the IFBB was evaluated using the range of data with the lowest standard deviation at the end of every operation stage. The influent (S_{in}) and effluent (S_{out}) COD and sulphate concentrations (mg.L^{-1}) were used to calculate the removal efficiencies (RE, Eq. 4-8), loading (LR, Eq. 4-9) and removal (RR, Eq. 4-10) rates. The Q_{in} was based on the period of operation and the IFBB volume (V_R) was constant at 2.5 L. The ratio between Q_{in} and V_R defines the hydraulic retention time (HRT). The total volatile suspended solids in the IFBB (X) was calculated using Eq. 4-11, which is according to the literature [26] where the authors did state that 83.4% of the VSS was determined as attached and the remainder was suspended.

$$RE = (S_{in} - S_{out}/S_{in}) \times 100 \qquad \text{Eq. 4-8}$$

$$LR = (Q/V_R) \times (S_{in}) \qquad \text{Eq. 4-9}$$

$$RR = (Q/V_R) \times (S_{in} - S_{out}) \qquad \text{Eq. 4-10}$$

$$Total\ biomass = X = (VSS)/0.166 \qquad \text{Eq. 4-11}$$

4.2.7   Chemical analysis

The chemical oxygen demand (COD, determined by the close reflux colorimetric method), VSS, TSS and total dissolved sulphide (sulphide or S^{2-}, determined by methylene blue reaction) were measured according to the standard methods for the examination of water and wastewater [27]. Sulphate was determined by ion chromatography (ICS-1000 Dionex, ASI-100 Dionex) as described by Villa-Gomez et al. [7].

Table 4-1. Schedule of IFBB operational parameters as applied during the different operational periods

Period of operation (symbol)	HRT (d)	COD$_{in}$ (mg.L^{-1})	SO$_4^{2-}$$_{in}$ (mg.L^{-1})	COD/SO$_4^{2-}$	pH$_{in}$	COD$_{out}$ (mg.L^{-1})	SO$_4^{2-}$$_{out}$ (mg.L^{-1})	Total dissolved sulphide (mg.L^{-1})	pH$_{out}$	VSS$_{out}$ (mg.L^{-1})	X (mg VSS.L^{-1})	X$_{rate}$ (mg VSS.L^{-1}d^{-1})
I (■)	1	1,142±205	712±42	1.6±0.2	6.2±0.2	369±63	584±34	42±29	7.2±0.17	303±55	1,827	1,827
II (◆)	0.5	912±63	752±20	1.2±0.1	6.0±0.1	436±35	523±31	122±22	7.0±0.03	435±285	2,620	5,241
III (*)	0.25	1,042±123	746±46	1.4±0.2	5.9±0.4	319±13	248±2	159±20	7.0±0.08	273±108	1,642	6,569
IV (●)	0.25	1,730±276	743±43	2.3±0.5	5.8±0.8	313±54	82±16	191±5	6.9±0.10	163±42	984	3,936
V (▲)	0.125	1,755±210	737±45	2.4±0.4	5.3±0.7	346±307	271±58	166±11	6.8±0.09	220±46	1,325	10,602
VI (○)	0.125	1,202±143	759±16	1.6±0.2	6.0±0.5	391±44	335±29	140±4	7.0±0.17	316±217	1,905	15,241
VII (□)	0.125	1,757±122	769±12	2.3±0.2	5.2±0.2	298±48	161±34	193±2	7.0±0.14	443±61	2,671	21,365
VIII (×)	0.125	1,212±147	743±59	1.7±0.1	5.2±0.1	195±58	250±42	164±18	7.1±0.12	255±220	1,536	12,289

4.2.8 Kinetic analysis

4.2.8.1 Second order substrate removal model

The Grau second order kinetic model [23] is described by Eq. 4-12, when this is integrated and linearized in the form of Eq. 4-13, the coefficient "a" is defined by Eq. 4-14. The coefficients "a" and "b" are calculated by linear regression of the resulting plot of the first term of Eq. 4-13 versus the HRT and further evaluated by the square of the Pearson correlation value (R^2). The second order ($k_{2(S)}$) rate constant is calculated after rearranging Eq. 4-14 into Eq. 4-15, wherein X_0 represents the biomass inside the IFBB. According to the research of Grau *et al.* [23], the coefficient "b" reflects the impossibility to reach zero values for the effluent substrate and remains close to one.

$$-dS/dt = k_{2(S)}X(S_{out}/S_{in})^2 \qquad \text{Eq. 4-12}$$

$$S_{in}.\,HRT/(S_{in} - S_{out}) = a + b.\,HRT \qquad \text{Eq. 4-13}$$

$$a = S_{in}/k_{2(S)}X_0 \qquad \text{Eq. 4-14}$$

$$k_{2(S)} = (1/a).\,(S_{in}/X_0) \qquad \text{Eq. 4-15}$$

4.2.8.2 The Stover-Kincannon model

The Stover-Kincannon model [24] is defined in Eq. 4-16 and describes the biological activity in a surface area (A). When the total biomass inside the V_R is taken into account, a modified model results as in Eq. 4-17. Substituting the substrate utilization rate (dS/dt) in Eq. 4-17 by the second term of Eq. 4-18 becomes Eq. 4-19. The constant of the maximum utilization rate (U_{max}) and the saturation constant (k_B) can be estimated after linearizing the Eq. 4-19 using the reciprocal and become the Eq. 4-20.

$$dS/dt = U_{max}(QS_{in}/A)/k_B + (QS_{in}/A) \qquad \text{Eq. 4-16}$$

$$dS/dt = U_{max}(QS_{in}/V_R)/k_B + (QS_{in}/V_R) \qquad \text{Eq. 4-17}$$

$$dS/dt = (Q/V_R)(S_{in} - S_{out}) \qquad \text{Eq. 4-18}$$

$$(Q/V_R)(S_{in} - S_{out}) = U_{max}(QS_{in}/V_R)/k_B + (QS_{in}/V_R) \qquad \text{Eq. 4-19}$$

$$V_R/Q\,(S_{in} - S_{out}) = k_B/U_{max}.\,V_R/QS_{in} + 1/U_{max} \qquad \text{Eq. 4-20}$$

4.3   Results

4.3.1   Hydrodynamic evaluation

4.3.1.1   Residence time distribution

The profile of the RTD in the IFB bioreactor is shown in Figure 4-1A. For the three recirculation flow velocities, the average mixing time was $\theta = 0.186 \pm 0.01$ (0.02325 d or 0.558 h); this was the time for the maximum signal of mass in the outlet of the reactor after the NaCl spike, $E(\theta)$ $= 0.84 \, (\pm 0.01)$. The fraction of mass $F(t) = 0.6 \, (\pm 0.001)$ left the reactor at $\theta = 1$. The maximum value of θ was 3.84 (± 0.013) and corresponded to $F(t) = 1$. For the three tests, the average τ_m was 3.12 (± 0.26) h (≈ 0.13 d), the recirculation flow velocity made a difference of 8.34% (Table 4-2).

4.3.1.2   Relative bed expansion

The RBE was plotted against the down flow liquid velocity used to expand the carrier material (Figure 4-1B). The "y" axis represented the vertical length or height of the column, zero was the top and the fraction 1 denoted the bottom. The horizontal line at fraction 0.8 represented the place of the outlet. There are two important aspects in the performance of the expansion: i) few carrier particles left the reactor through the outlet when the bed was expanded to the RBE = 0.8 and ii) the carrier material was not expanded to the maximum RBE (1.0), because they were sucked out by the recirculation pump and damaged the fluidization system. Hence, a RBE = 0.5 was selected to avoid the suction by the recirculation pump and washout of carrier material from the IFBB during its operation (155 d).

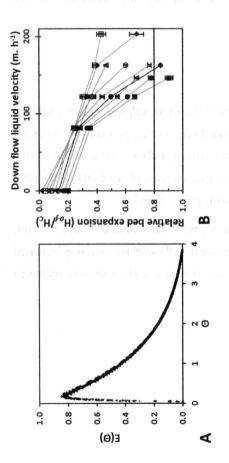

Figure 4-1. Hydrodynamic characterization of the inversed fluidized bed bioreactor

A) Residence time distribution at three different down flow liquid velocities, 81 m. h^{-1} ($\Diamond$), 122 m. h^{-1} or ($\triangle$) and 146 m. h^{-1} or (+). B) Relative bed expansion in the column. Different initial volumes of carrier material or static bed were tested: 50 mL ($\blacklozenge$), 100 mL ($\square$), 150 mL ($\triangle$), 200 mL ($\circ$), 250 mL ($\times$), 300 mL ($\Diamond$), 350 mL ($\blacktriangle$), 400 mL ($\bullet$), and 450 mL ($\blacksquare$) at different down flow liquid velocities, ranging from 0-203 m. h^{-1} (recirculation flow velocity 0-500 L. h^{-1}).

Table 4-2. Summary of RTD analysis at different recirculation flow velocities

Recirculation flow velocity	Θ	$E(\Theta)$	τ_m	Θ at F(0.5)	F(t) at Θ=1	Θ_{max} at F(t)=1	V_R (L)	Variance (σ^2)	SD (σ)
200 L.h^{-1}	0.177	0.85	3.30	0.804	0.594	3.82	2.75	6.25	2.50
300 L.h^{-1}	0.180	0.84	3.25	0.801	0.595	3.85	2.71	6.13	2.48
360 L.h^{-1}	0.201	0.82	2.82	0.803	0.596	3.84	2.35	4.59	2.14
Average	0.186	0.84	3.12	0.803	0.595	3.84	2.60	5.66	2.37
SD	0.013	0.01	0.26	0.002	0.001	0.01	0.22	0.93	0.2
error (%)	7.0	1.5	8.3	0.2	0.1	0.3	8.3	16.4	8.4

The volumes 50, 100, 150, 200, 250, 300, 350, 400 and 450 mL of carrier material (H_0) produced, without recirculation flow, a RBE also named relative static bed (RSB) equal to 0.022, 0.045, 0.065, 0.085, 0.11, 0.135, 0.15, 0.175 and 0.2 (RBE and RSB are dimensionless) respectively. The RBE was linear using a RSB < 0.065 (H_0 < 150 mL of carrier material) when operated at a DFLV = 0-162 m.h^{-1} (RFV = 0-400 L.h^{-1} of) and beads were not expanded > 40% of the column length.

A relative static bed (RSB) within the range ≥ 0.085 and ≤ 0.15 (H_0 ≥ 200 to H_0 ≤ 350 mL of carrier material) showed resistance to fluidize at a DFLV ≤ 81 m.h^{-1} (RFV ≤ 200 L.h^{-1}) but fluidized longer in comparison to lower RSB (≤ 0.085) at DFLV > 81 m.h^{-1} (RFV > 200 L.h^{-1}). The RSB between the range 0.085-0.15 could not be tested at DFLV > 162 m.h^{-1} (RFV > 400 L.h^{-1}): the carrier material was washed out from the IFBB. The RBE was larger using RSB ≥ 0.175 and ≤ 0.2 and the expansion was affected exponentially by the DFLV tested (0-146 m.h^{-1}). At a DFLV > 146 m.h^{-1} (RFV > 360 L.h^{-1}), the carrier material was washed out from the IFBB and, therefore, higher DFLV were not tested.

4.3.2 Sulphate reduction in the high rate IFBB

4.3.2.1 Sulphate and COD removal efficiency

Figure 4-2A-E shows the profiles of COD and sulphate RE, COD, sulphate, sulphide effluent concentrations and the influent and effluent pH, respectively. During period I (6 d duration) and an HRT = 1 d, the COD and sulphate RE was, respectively, 68% and 18%. Period II (10 d duration) was operated at an HRT = 0.5 d, resulting in a decrease of the COD RE to 52%, whereas the sulphate RE increased 30% compared to the previous period.

In period III, the HRT was decreased to 0.25 d; the RE was improved to 69% and 67% for, respectively, COD and sulphate. In period IV, the influent COD:sulphate ratio was increased from 1.4 to 2.3 compared to the previous period III. The COD and sulphate RE were improved to 82% and 89%, respectively.

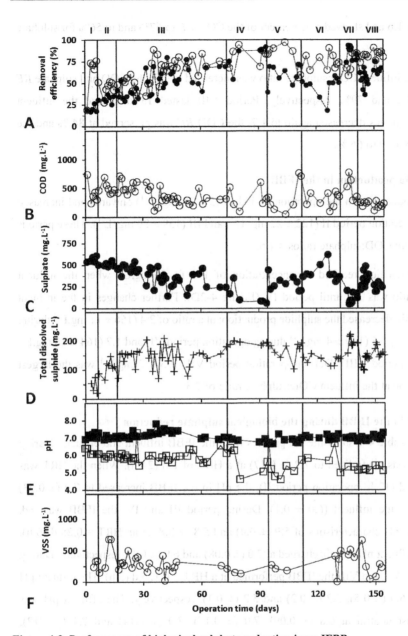

Figure 4-2. Performance of biological sulphate reduction in an IFBB
A) Sulphate (●) and COD (○) removal efficiency; B) COD concentration in the effluent (○); C) Sulphate concentration in the effluent (●); D) Sulphide concentration in the effluent (+); E) Influent (□) and effluent (■) pH; F) VSS concentration in the effluent (○)

In period V, the HRT was decreased further to 0.125 d (3 h), the COD *RE* was maintained at 80% and the sulphate *RE* decreased to 63%. During period VI, the influent COD:sulphate ratio was decreased to 1.6 and showed a decrement of the COD *RE* to 67% and to 56% for sulphate *RE*.

In period VII, the influent COD:sulphate ratio was increased to 2.3; the COD and sulphate *RE* improved to 83% and 79%, respectively. Period VIII lasted 13 days and the influent COD:sulphate ratio was decreased again to 1.7; the COD *RE* was preserved at 84 % and the sulphate *RE* decreased to 66 %.

4.3.2.2 Sulphide production in the IFBB

Sulphide was produced from the operation period I (42 ± 29 mg. L^{-1}) onwards and increased throughout the operation period II (122 ± 22 mg. L^{-1}) and III (159 ± 20 mg. L^{-1}). These periods operated at influent COD:sulphate ratios < 1.6.

The sulphide production reached a concentration of 191 (± 5) mg.L^{-1} when the influent COD:sulphate ratio was 2.3 until period IV (Figure 4-2D). Further changes in the influent COD:sulphate ratio decreased the sulphide production: at a ratio of 2.4 (166 ± 11 mg.L^{-1} during operation period V), 1.6 (140 ± 4 mg.L^{-1} during operation period VI) and 1.7 (164 ± 18 mg.L^{-1} during operation period VIII). During operation period VII, 193 ± 2 mg.L^{-1} was the largest sulphide production at the influent COD:sulphate ratio of 2.3.

4.3.2.3 The pH in the IFBB during the biological sulphate reduction

Figure 4-2E and Table 4-1 show the pH profiles from the IFBB influent and effluent. During period I, the pH was neutralized to 7.2 (± 0.17) at a HRT of 1 d (24 h). When the HRT was decreased to 0.5 d (12 h, operation period II), the pH in the IFBB increased to 7.0 (± 0.03) despite the pH of the influent (6.0 $\pm$ 0.1). During period III and IV, the IFBB was fed, respectively, with pH characteristics of 5.9 (± 0.4) and 5.8 (± 0.8) at an HRT = 0.25 d (6 h). Meanwhile the effluent pH was discharged at 7.0 (± 0.08) and 6.9 (± 0.1), respectively. During the periods V, VI, VII and VIII the IFBB performed at a HRT = 0.125 d (3 h). The influent pH was 5.3 (± 0.7), 6.0 (± 0.5), 5.2 (± 0.2) and 5.2 (± 0.1), respectively. The effluent pH was maintained almost neutral at 6.8 (± 0.09), 7.0 (± 0.17), 7.0 (± 0.14) and 7.1 (± 0.12), respectively, in the IFBB.

4.3.2.4 Biomass production during the IFBB operation

The initial biomass concentration was 1,610 mg VSS.L^{-1} (1.61 g VSS.L^{-1}) in the IFBB, after the mixing of the inoculum (0.25 L at 16.1 g VSS.L^{-1}) in the IFBB volume. There was a continuous production of VSS that averaged 301 (± 98) mg VSS.L^{-1} in the IFBB effluent

(Figure 4-2F). According to Eq. 4-11, the calculated total biomass in the IFBB (X) averaged 1,814 ($\pm$ 589) mg VSS.L^{-1}. Table 4-1 and Figure 4-3 shows the volumetric production rate of biomass (X_{rate}). The IFBB was stopped after 155 days of operation, the final TSS and VSS concentrations accumulated were 4,767 ($\pm$ 386) and 1,600 ($\pm$ 96) mg.L^{-1}, respectively.

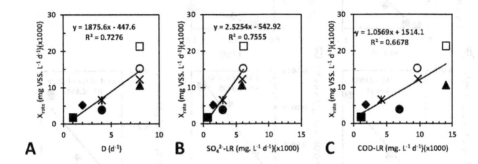

Figure 4-3. VSS production rate in the IFBB (X_{rate})
The X_{rate} against the dilution rate (A), sulphate (B) and COD (C) loading rate. The period I ($\blacksquare$), II ($\blacklozenge$), III (*), IV ($\bullet$), V ($\blacktriangle$), VI ($\circ$), VII ($\square$) and VIII ($\times$)

4.3.3 Kinetic analysis of the IFBB performance

4.3.3.1 Grau second order substrate removal

The Grau second order (Eq. 4-13) approach showed great values of correlation ($R^2 > 0.95$) in the case of sulphate and COD removal (Figure 4-4A-B). In this research, constant "b" was 6 for the sulphate removed and ≈ 1.6 for the simultaneous COD removal. The constant "a" (Eq. 4-14) was -0.799 for sulphate removed and -0.025 COD removal. The $k_{2(S)}$ was calculated using Eq. 4-15, the VSS at t = 155 d (1,600 $\pm$ 96 mg VSS.L^{-1}), the average sulphate (745 $\pm$ 17 mg. L^{-1}) and COD concentration (1,344 $\pm$ 347 mg. L^{-1}) after 155 d of operation (Table 4-3). The calculated $k_{2(S)}$ (d^{-1}) was, respectively, -0.58 and -33.92 for sulphate and COD removal.

4.3.3.2 The Stover-Kincannon model

The Stover-Kincannon model (Eq. 4-17) did fit to the experimental data of the IFBB operation with $R^2 \geq 0.89$ (Figure 4-4C-E), therefore, k_B and the U_{max} were calculated using Eq. 4-20. The respective U_{max} rate constant was 934 mg SO_4^{2-}.L^{-1}d^{-1} and 45,669 mg COD.L^{-1}d^{-1}. The respective k_B saturation rate constant was 5,609 mg SO_4^{2-}.L^{-1}d^{-1} and 73,949 mg COD.L^{-1}d^{-1} in the IFBB (Table 4-3).

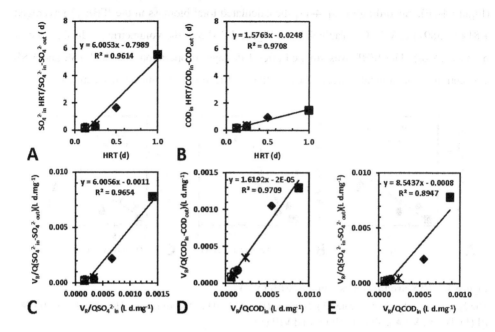

Figure 4-4. Kinetic evaluation of biological sulphate reduction in an anaerobic IFBB
Grau second order substrate removal (A-B), Stover-Kincannon (C-E). Period I (■), II (◆), III (＊), IV (●), V (▲), VI (○), VII (□) and VIII (×)

Table 4-3. Kinetic constants of the different substrate removal models used to analyze the kinetics of the IFBB

	SO_4^{2-}	COD	SO_4^{2-} on COD
Grau Second order			
$k_{2(S)}$ (d^{-1})	-0.58	-33.92	
a	-0.799	-0.025	
b	6.005	1.576	
Stover-Kincannon			
U_{max} (mg.L^{-1}d^{-1})	934	45,669	1,222
K_B (mg.L^{-1}d^{-1})	5,609	73,949	10,438

4.4   Discussion

4.4.1   IFBB hydrodynamic performance

This study showed that biological sulphate reduction is possible at a HRT = 0.125 d. Technical issues take place during the long term operation of the IFBB, *e.g.* the agglomeration of the carrier due to biofilm on the carrier surface and further precipitation because of the increasing bed density. The suction of the carrier material gave operational problems in the recirculation pump [14], this is also possible at an $RBE = 1$ only with the empty carrier beads (Figure 4-1B).

The minimum fluidization velocity was reported as 0.0083 $m.s^{-1}$ (= 30 $m.h^{-1}$) for low density polyethylene beads with spherical shape and 4 mm diameter in a downwards liquid fluidized column with a H/D ratio of 19.1 [28], apparently these values were calculated within the RSB corresponding to $H_0/H_C \approx 0.016\text{-}0.05$, in this range there was no variation in the minimum fluidization velocity. In contrast, with a longer RSB ($H_0/H_C) \geq 0.08$ and polyethylene as carrier, the minimum fluidization velocity is positively affected: a larger RBE at a lower DFLV can be achieved. Therefore, less energy has to be consumed in order to fluidize the carrier material [9,14]. However, DFLV has a larger effect than increasing RSB on the bead expansion [14].

On the other hand, the hydrodynamic properties illustrated by the RTD (Figure 4-1A) were not negatively affected at the RFV (200, 300 and 360 $L.h^{-1}$) tested on the IFBB. This was shown when τ_m (3.12 ± 0.26 h or ≈ 0.13 d) was measured at the three RFV and the standard deviation represented an error of 8.34% (Table 4-2). The liquid velocity improves the axial dispersion and subsequently the mixing capacity of the reactor [9]. This suggested that the RFV (200, 300 and 360 $L.h^{-1}$) tested did not affected the mixing times ($\theta = 0.186 \pm 0.01$ equivalent to 0.558 h) in the IFBB. Furthermore, according to the literature [29] hydrodynamic characteristic of non-ideal CSTR are long tails expressed as large values for maximum retention time. The maximum retention time observed ($\theta = 3.84 \pm 0.013$ at $F(t) = 1$) suggested that the IFBB performed hydrodynamically like a non-ideal CSTR.

4.4.2 Biomass retention in the IFBB

The uncoupling of the biomass retention from the liquid retention is an advantage when IFBB are used to perform biological sulphate reduction at high rate (HRT = 0.125 d) conditions. In this research, the IFBB did not use biomass recirculation. The absence of biomass recirculation favoured the attachment of biomass on a carrier bed and outcompeted suspended cells [30,31]. This is possible by controlling certain cell groups due to the relationship of the cell specific growth rate (μ with units of time^{-1}) and the hydrodynamic conditions imposed by the dilution, D or HRT^{-1} as in Eq. 4-21 with units of time^{-1}.

$$\mu = D = 1/HRT \hspace{4cm} \text{Eq. 4-21}$$

Additionally, the presence of SO_4^{2-} in the inorganic wastewater suggests the deconvolution of the cell retention from the liquid retention time. SO_4^{2-} has stronger ionic interactions than other ions with water, *e.g.* Cl⁻ [32]. This can destabilise the already weak interaction of cell-H-O-H-cell, which is also supported by the low negative charges and hydrophobicity of cells [33].

Three major groups dominate sulphate reducing consortia [34,35]: hydrolytic fermentative bacteria, SRB, and methanogenic archaea, these have different specific growth rates that can be affected by the HRT (Eq. 4-21). However, in the IFBB, the X_{rate} (mg VSS.L^{-1}d^{-1}) was not affected and increased simultaneously with the D (HRT^{-1} as in Eq. 4-21) at a concentration of 1,875 mg.L^{-1} (R^2 = 0.72, Figure 4-3A). Also, the generation rate of VSS in the IFBB (X_{rate}) was supported by the θ = 3.84 ($\pm$ 0.013) at $F(t)$ = 1.

Furthermore, the electron donor (mg COD$_{Lactate}$.L^{-1}d^{-1}) and sulphate (mg SO$_4^{2-}$.L^{-1}d^{-1}) loading rate did positively influence the X_{rate}, this is observed in the positive slopes indicating the generation of new VSS on the sulphate and COD consumption, respectively, 2.52 mg VSS.mg [SO$_4^{2-}$]$^{-1}$ (Figure 4-3B) and 1.05 mg VSS.mg COD^{-1} (Figure 4-3C). This VSS yield on sulphate was 15 times smaller in comparison to that reported for a membrane bioreactor performing sulphate reduction at pH 6 (38 mg VSS. mg [SO$_4^{2-}$]$^{-1}$) using formate as electron donor in a pH auxostat system [36].

4.4.3 Sulphate reduction in the IFBB at 3 h HRT

The IFBB 3 h hydrodynamic parameters from this research (two phases solid-liquid) shared characteristics, *e.g.* mixing ($\theta \approx 0.2$) and maximum retention time ($\theta > 3.7$ and $\theta < 4.0$), with three phases IFBB (solid-liquid-gas) [37] and with membrane bioreactors [38]. In principle, UASB bioreactors are well known for their robustness linked to high biomass retention in the form of anaerobic granules and membrane bioreactors for the separation of biomass from the liquid by a synthetic barrier. Furthermore, high sulphate RE using an IFBB is comparable to processes using UASB and membrane bioreactors (75-85 %) at a HRT = 15.5 h [39].

The VSS concentration (X_0 = VSS) can determine the bioreactor kinetics order during anaerobic processes [40]. Additionally, biological sulphate removal processes are hampered if the COD concentration is limiting [41]. High sulphate RR within a magnitude of 4,866 mg SO$_4^{2-}$.L^{-1}d^{-1} (period VII) was associated to the biomass (X) concentration (mg VSS.L^{-1}) during the 155 d of IFBB operation (Table 4-1). According to Grau *et al.* [23], the biological reaction kinetics are influenced by S_{in}/X_0 ratio as in Eq. 4-15, therefore, a small S_{in}/X_0 ratio is influenced by a large X concentration (mg VSS.L^{-1}). Using the IFBB, the Grau second order model fitted with $R^2 >$ 0.95 in the case of sulphate and COD removal (Figure 4-4A-B) and showed that the biomass in the reactor was not limiting the sulphate removal. Bioreactors operated at small S_{in}/X_0 ratios, building block anabolic pathways are preferred rather than catabolic pathways [42]. The $k_{2(S)}$ constant of sulphate removal was tightly linked to the X_{rate} and therefore linked to the COD: sulphate ratio of 2.3.

On the other hand, the Stover-Kincannon U_{max} (934 mg SO_4^{2-}. L^{-1} d^{-1}) constant (Table 4-3) was superior to that reported for the propionate use in a sequencing batch reactor operating at an HRT of 2 d (700 mg $SO_4^{2-}.L^{-1}d^{-1}$) [43]. The Stover-Kincannon k_B (5,609 mg $SO_4^{2-}.L^{-1}d^{-1}$) constant was slightly larger when compared to the RR of 4,866 mg $SO_4^{2-}.L^{-1}d^{-1}$ observed using lactate as electron donor, COD: sulphate ratio of 2.3, pH of 5.2 and HRT of 0.125, during period VII. A RR of 525 mg $SO_4^{2-}.L^{-1}d^{-1}$ was observed (pH = 5.24 and an HRT = 1 d) using limiting lactate conditions and COD: sulphate ratio of 0.67 in an IFBB [44]. Moreover, Papirio et al [44] reported 63 % of sulphate RE for a sulphate LR of 1000 mg $SO_4^{2-}.L^{-1}d^{-1}$, this is 630 mg $SO_4^{2-}.L^{-1}d^{-1}$ (pH = 5.31 and an HRT = 1 d), using lactate at a COD:sulphate ratio of 4 in an IFBB. In the literature [43,44], the sulphate RR were affected by the HRT used (1 d), this suggests that short HRT are beneficial for biological sulphate removal.

First order substrate removal and Monod models could not be applied in this research (result are not shown, $R^2 < 0.65$), despite these models are extensively applied to describe the sulphate reduction mediated by SRB in chemostat conditions [19,20]. The Grau second order and Stover-Kincannon models have been used to evaluate the reactor performance with high biomass retention and COD removal [21,22], but not for sulphate removal in an IFBB. The Stover-Kincannon model has theoretically supported the observed rates (4,866 mg $SO_4^{2-}.L^{-1}d^{-1}$) during the IFBB operation under the conditions tested (Table 4-1). Also, the Grau second order substrate removal model confirmed the effect of biomass concentration in the IFBB (X) during the performance of sulphate removal.

During the IFBB operation, the produced sulphide concentrations ranged from 42 ($\pm$ 29) to 193 ($\pm$ 2) mg TDS.L^{-1}, such concentrations did not represent a risk to inhibit the system. Bioreactors like the IFBB using anaerobic sludge as source of SRB can operate at 1,200 mg.L^{-1} of sulphide without affecting the COD and sulphate RE [45]. On the other hand, the potential toxicity of sulphide can be reduced simultaneously to the decreasing HRT [46].

4.5   Conclusions

An IFBB showed capability to perform biological sulphate reduction and biomass retention at an HRT as low as 0.125 d (3 h), this was supported by the maximum retention time ($\theta = 3.84 \pm 0.013$) measured for the IFBB. The sulphate RR (4,866 mg SO_4^{2-}. L^{-1} d^{-1} at HRT=0.125 d and

COD: $SO_4^{2-}= 2.3$) was not limited by the decrease of the HRT during the 155 d of operation. The IFBB performance was influenced by the COD:sulphate ratio, the best sulphate RR were observed during periods IV and VII at a COD:SO_4^{2-}ratio of 2.3, respectively at an HRT as low as 0.25 and 0.125 d. The bioprocess of anaerobic sulphate reduction in the IFBB was robust under all conditions tested. The Grau second order and Stover-Kincannon substrate removal model appropriately fitted to the experimental data ($R^2 > 0.96$). The models supported that biomass was not a limiting conditions and the sulphate RR at HRT as low as 3 h in the IFBB was optimal for sulphate reduction.

4.6   References

[1]   C. Monterroso, F. Macías, Drainage waters affected by pyrite oxidation in a coal mine in Galicia (NW Spain): Composition and mineral stability, Sci. Total Environ. 216 (1998) 121–132. doi:10.1016/S0048-9697(98)00149-1.

[2]   H. Bai, Y. Kang, H. Quan, Y. Han, J. Sun, Y. Feng, Treatment of acid mine drainage by sulfate reducing bacteria with iron in bench scale runs, Bioresour. Technol. 128 (2013) 818–822. doi:10.1016/j.biortech.2012.10.070.

[3]   A.M. Sarmiento, M. Olías, J.M. Nieto, C.R. Cánovas, J. Delgado, Natural attenuation processes in two water reservoirs receiving acid mine drainage, Sci. Total Environ. 407 (2009) 2051–2062. doi:10.1016/j.scitotenv.2008.11.011.

[4]   F. Mapanda, G. Nyamadzawo, J. Nyamangara, M. Wuta, Effects of discharging acid-mine drainage into evaporation ponds lined with clay on chemical quality of the surrounding soil   and   water,   Phys.   Chem.   Earth,   Parts   A/B/C.   32   (2007)   1366–1375. doi:10.1016/j.pce.2007.07.041.

[5]   N.S. Cheung, Z.F. Peng, M.J. Chen, P.K. Moore, M. Whiteman, Hydrogen sulfide induced neuronal death occurs via glutamate receptor and is associated with calpain activation and lysosomal rupture in mouse primary cortical neurons, Neuropharmacology. 53 (2007) 505–514. doi:10.1016/j.neuropharm.2007.06.014.

[6]   R. Jachuck, J. Lee, D. Kolokotsa, C. Ramshaw, P. Valachis, S. Yanniotis, Process intensification for energy saving, Appl. Therm. Eng. 17 (1997) 861–867. doi:10.1016/S1359-4311(96)00048-8.

[7]   D. Villa-Gomez, H. Ababneh, S. Papirio, D.P.L. Rousseau, P.N.L. Lens, Effect of sulfide concentration on the location of the metal precipitates in inversed fluidized bed reactors, J. Hazard. Mater. 192 (2011) 200–207. doi:10.1016/j.jhazmat.2011.05.002.

[8]   D. Prato-Garcia, F.J. Cervantes, G. Buitrón, Azo dye decolorization assisted by chemical and biogenic sulfide, J. Hazard. Mater. 250–251 (2013) 462–468. doi:10.1016/j.jhazmat.2013.02.025.

[9]   T. Renganathan, K. Krishnaiah, Liquid phase mixing in 2-phase liquid–solid inverse fluidized bed, Chem. Eng. J. 98 (2004) 213–218. doi:10.1016/j.cej.2003.08.001.

[10]   M. Comte, D. Bastoul, G. Hebrard, M. Roustan, V. Lazarova, Hydrodynamics of a three-phase fluidized bed—the inverse turbulent bed, Chem. Eng. Sci. 52 (1997) 3971–3977. doi:10.1016/S0009-2509(97)00240-6.

[11]   P. Buffière, R. Moletta, Some hydrodynamic characteristics of inverse three phase fluidized-bed reactors, Chem. Eng. Sci. 54 (1999) 1233–1242. doi:10.1016/S0009-2509(98)00436-9.

[12]   M. Rajasimman, C. Karthikeyan, Aerobic digestion of starch wastewater in a fluidized bed bioreactor with low density biomass support, J. Hazard. Mater. 143 (2007) 82–86. doi:10.1016/j.jhazmat.2006.08.071.

[13]   D. Garcia-Calderon, P. Buffiere, R. Moletta, S. Elmaleh, Anaerobic digestion of wine distillery wastewater in down-flow fluidized bed, Water Res. 32 (1998) 3593–3600. doi:10.1016/S0043-1354(98)00134-1.

[14]   A. Alvarado-Lassman, E. Rustrián, M.A. García-Alvarado, G.C. Rodríguez-Jiménez, E. Houbron, Brewery wastewater treatment using anaerobic inverse fluidized bed reactors, Bioresour. Technol. 99 (2008) 3009–3015. doi:10.1016/j.biortech.2007.06.022.

[15]   A. Alvarado-Lassman, E. Rustrián, M.A. García-Alvarado, E. Houbron, Simultaneous removal of carbon and nitrogen in an anaerobic inverse fluidized bed reactor, Water Sci. Technol. 54 (2006) 111–117. doi:10.2166/wst.2006.493.

[16]   P. Kijjanapanich, A.T. Do, A.P. Annachhatre, G. Esposito, D.H. Yeh, P.N.L. Lens, Biological sulfate removal from construction and demolition debris leachate: Effect of bioreactor configuration, J. Hazard. Mater. 269 (2014) 38–44. doi:10.1016/j.jhazmat.2013.10.015.

[17]   L.C. Reyes-Alvarado, N.N. Okpalanze, D. Kankanala, E.R. Rene, G. Esposito, P.N.L. Lens, Forecasting the effect of feast and famine conditions on biological sulphate reduction in an anaerobic inverse fluidized bed reactor using artificial neural networks, Process Biochem. 55 (2017) 146–161. doi:10.1016/j.procbio.2017.01.021.

[18]   S.L. Ong, A comparison of estimates of kinetic constants for a suspended growth treatment system from various linear transformations, Res. J. Water Pollut. Control Fed. 62 (1990) 894–900.

[19]   W.-C. Kuo, T.-Y. Shu, Biological pre-treatment of wastewater containing sulfate using anaerobic immobilized cells, J. Hazard. Mater. 113 (2004) 147–155. doi:10.1016/j.jhazmat.2004.05.033.

[20]   S. Moosa, M. Nemati, S.T.L. Harrison, A kinetic study on anaerobic reduction of sulphate, Part I: Effect of sulphate concentration, Chem. Eng. Sci. 57 (2002) 2773–2780. doi:10.1016/S0009-2509(02)00152-5.

[21]   M. Işik, D.T. Sponza, Substrate removal kinetics in an upflow anaerobic sludge blanket reactor decolorising simulated textile wastewater, Process Biochem. 40 (2005) 1189–1198. doi:10.1016/j.procbio.2004.04.014.

[22]   M. Pandian, H. Ngo, S. Pazhaniappan, Substrate removal kinetics of an anaerobic hybrid reactor treating pharmaceutical wastewater, J. Water Sustain. 1 (2011) 301–312.

[23]   P. Grau, M. Dohányos, J. Chudoba, Kinetics of multicomponent substrate removal by activated sludge, Water Res. 9 (1975) 637–642. doi:10.1016/0043-1354(75)90169-4.

[24]   E.L. Stover, D.F. Kincannon, Evaluating rotating biological contactor performance, Water Sew. Work. 123 (1976) 88–91.

[25]   H. Scott Fogler, Elements of chemical reaction engineering, 3rd ed., Upper Saddle River, N.J. : Prentice Hall PTR, 1987. doi:10.1016/0009-2509(87)80130-6.

[26]   C. Sheli, R. Moletta, Anaerobic treatment of vinasses by a sequentially mixed moving bed biofilm reactor, Water Sci. Technol. 56 (2007) 1–7. doi:10.2166/wst.2007.465.

[27]   APHA, Standard Methods for the Examination of Water and Wastewater, 20th ed., Washington, USA., 1999.

[28]   A.C. Vijaya Lakshmi, M. Balamurugan, M. Sivakumar, T. Newton Samuel, M. Velan, Minimum fluidization velocity and friction factor in a liquid-solid inverse fluidized bed reactor, Bioprocess Eng. 22 (2000) 461–466. doi:10.1007/s004490050759.

[29]   A. Martin, Interpretation of residence time distribution data, Chem. Eng. Sci. 55 (2000) 5907–5917. doi:10.1016/S0009-2509(00)00108-1.

[30]   R. Cresson, R. Escudié, J.P. Steyer, J.P. Delgenès, N. Bernet, Competition between planktonic and fixed microorganisms during the start-up of methanogenic biofilm reactors, Water Res. 42 (2008) 792–800. doi:10.1016/j.watres.2007.08.013.

[31]   R. Escudié, R. Cresson, J.-P. Delgenès, N. Bernet, Control of start-up and operation of anaerobic biofilm reactors: An overview of 15 years of research, Water Res. 45 (2011) 1–10. doi:10.1016/j.watres.2010.07.081.

[32]   B. Hribar, N.T. Southall, V. Vlachy, K.A. Dill, How Ions Affect the Structure of Water, J. Am. Chem. Soc. 124 (2002) 12302–12311. doi:10.1021/ja026014h.

[33]   J.S. Dickson, M. Koohmaraie, Cell surface charge characteristics and their relationship to bacterial attachment to meat surfaces, Appl. Environ. Microbiol. 55 (1989) 832–836.

[34]   J. Wang, M. Shi, H. Lu, D. Wu, M.-F. Shao, T. Zhang, G.A. Ekama, M.C.M. van Loosdrecht, G.-H. Chen, Microbial community of sulfate-reducing up-flow sludge bed in the SANI® process for saline sewage treatment, Appl. Microbiol. Biotechnol. 90 (2011) 2015–2025. doi:10.1007/s00253-011-3217-3.

[35]   S.J.W. Oude Elferink, W.J. Vorstman, A. Sopjes, A.J. Stams, Characterization of the sulfate-reducing and syntrophic population in granular sludge from a full-scale anaerobic reactor treating papermill wastewater, FEMS Microbiol. Ecol. 27 (1998) 185–194. doi:10.1111/j.1574-6941.1998.tb00536.x.

[36]   M.F.M. Bijmans, E. de Vries, C.-H. Yang, C.J. N. Buisman, P.N.L. Lens, M. Dopson, Sulfate reduction at pH 4.0 for treatment of process and wastewaters, Biotechnol. Prog. 26 (2010) 1029–1037. doi:10.1002/btpr.400.

[37]   O. Sánchez, S. Michaud, R. Escudié, J.-P. Delgenès, N. Bernet, Liquid mixing and gas–liquid mass transfer in a three-phase inverse turbulent bed reactor, Chem. Eng. J. 114 (2005) 1–7. doi:10.1016/j.cej.2005.08.009.

[38] M.W.D. Brannock, Y. Wang, G. Leslie, Evaluation of full-scale membrane bioreactor mixing performance and the effect of membrane configuration, J. Memb. Sci. 350 (2010) 101–108. doi:10.1016/j.memsci.2009.12.016.

[39] P. Kijjanapanich, A.T. Do, A.P. Annachhatre, G. Esposito, D.H. Yeh, P.N.L. Lens, Biological sulfate removal from construction and demolition debris leachate: Effect of bioreactor configuration, J. Hazard. Mater. 269 (2014) 38–44. doi:10.1016/j.jhazmat.2013.10.015.

[40] L. Neves, R. Oliveira, M.M. Alves, Influence of inoculum activity on the bio-methanization of a kitchen waste under different waste/inoculum ratios, Process Biochem. 39 (2004) 2019–2024. doi:10.1016/j.procbio.2003.10.002.

[41] A. Velasco, M. Ramírez, T. Volke-Sepúlveda, A. González-Sánchez, S. Revah, Evaluation of feed COD/sulfate ratio as a control criterion for the biological hydrogen sulfide production and lead precipitation, J. Hazard. Mater. 151 (2008) 407–413. doi:10.1016/j.jhazmat.2007.06.004.

[42] Y. Liu, G.-H. Chen, E. Paul, Effect of the S_0/X_0 ratio on energy uncoupling in substrate-sufficient batch culture of activated sludge, Water Res. 32 (1998) 2883–2888. doi:10.1016/S0043-1354(98)00071-2.

[43] R. Ghigliazza, A. Lodi, M. Rovatti, Kinetic and process considerations on biological reduction of soluble and scarcely soluble sulfates, Resour. Conserv. Recycl. 29 (2000) 181–194. doi:10.1016/S0921-3449(99)00055-5.

[44] S. Papirio, G. Esposito, F. Pirozzi, Biological inverse fluidized-bed reactors for the treatment of low pH- and sulphate-containing wastewaters under different COD conditions, Environ. Technol. 34 (2013) 1141–1149. doi:10.1080/09593330.2012.737864.

[45] L.B. Celis-García, E. Razo-Flores, O. Monroy, Performance of a down-flow fluidized bed reactor under sulfate reduction conditions using volatile fatty acids as electron donors, Biotechnol. Bioeng. 97 (2007) 771–779. doi:10.1002/bit.21288.

[46] A.H. Kaksonen, P.D. Franzmann, J.A. Puhakka, Effects of hydraulic retention time and sulfide toxicity on ethanol and acetate oxidation in sulfate-reducing metal-precipitating fluidized-bed reactor, Biotechnol. Bioeng. 86 (2004) 332–343. doi:10.1002/bit.20061.

Chapter 5

Effect of the initial sulphate concentration on the start-up phase of the biological sulphate reduction in sequencing batch reactors

Abstract

The influence of the initial sulphate concentration was investigated for time reduction of start-up phase on the sulphate removal process in sequencing batch reactors (SBR). Two SBR, named L and H, were operated with an influent sulphate concentration of 0.4 and 2.5 g SO_4^{2-} .L^{-1}, respectively. Lactate was used as electron donor at a COD: SO_4^{2-} ratio of 2.4 for 34 d. The SBR L was operated at a constant hydraulic retention time (HRT = 2 d) and volumetric feeding rate was disturbed for the SBR H (HRT = 2 d for 8 d, batch conditions for 6 d and HRT = 2 d for 20 d). The control reactor L had a lag phase for the sulphate removal efficiency (22 ± 15%) of 12 d and reached steady state conditions (90 ± 9%). The reactor H showed different sulphate removal efficiencies: lag phase (62 ± 25%) 8 d, batch (95 ± 4%) 6 d, SBR non steady state (65 ± 12%) 6 d and SBR steady state (96 ± 10%) 14 d. The SBR H showed higher sulphate removal efficiencies and the start-up phase was optimized at 2.5 g SO_4^{2-}.L^{-1}.

Keywords: Sulphate reduction, wastewater treatment, sequencing batch reactor, steady and non-steady state behaviour

5.1 Introduction

Industrial wastewater containing high sulphate concentrations represent an environmental risk, as this type of wastewater can dramatically damage the flora and fauna of water reservoirs (Mapanda *et al.*, 2007). Industrial wastewater with high sulphate concentrations are characterized by low pH, high oxidative potential, contains high concentrations of toxic metal and lack of chemical oxygen demand (COD). This last characteristic is an economic disadvantage for the biological treatment, as electron donors need to be supplied to fuel the sulphate reducing bacteria (SRB). Pure chemicals can be added as electron donors (COD), but this increases the overall costs of the biological process.

The removal of sulphate by SRB has been studied in different anaerobic reactor as *e.g.* batch reactor (Al-Zuhair *et al.*, 2008), sequencing batch reactor (Torner-Morales and Buitrón, 2010), upflow anaerobic sludge blanket reactor, (UASB) (Bertolino *et al.*, 2012), extended granular sludge bed reactor (EGSB) (Dries *et al.*, 1998) and gas lift reactor (Sipma *et al.*, 2007). Among this studies, the COD:sulphate ratio and the hydrodynamic conditions (effect of the HRT) have been studied extensively for high sulphate removal efficiencies. Nevertheless, conditions that might hamper or shorten the start-up phase of biological sulphate reduction are not well known.

For example, methanogenic pathways are optimized by addition of trace metals (as iron, cobalt and nickel) and sulphur compounds. The addition of iron showed a significant effect on the methane production using methanol as substrate in batch bioreactors (Zandvoort *et al.*, 2003). The supply of cysteine increased the trace metal retention time (after 6 d of application) and therefore the methanogenic activity (using methanol as substrate) showed an optimal activity in a UASB after 43 d of operation (Zandvoort *et al.*, 2005). Earlier, $FeSO_4$ was added to decrease the redox potential, decreased by means of biological sulphate reduction and the produced FeS, in a methanogenic UASB degrading 4-methylbenzoic acid (Macarie and Guyot, 1995).

Shortening the start-up phase can be also beneficial to decrease the cost of operation. For instance, the specific growth rates of pure cultures of SRB are stimulated when the concentration of sulphate reaches as high as 2.5 g $SO_4^{2-}.L^{-1}$ (Al-Zuhair *et al.*, 2008). It is unknown if the same effect can take place using a consortium such as anaerobic sludge that is widely used as inoculum of SRB during start-up phase in bioreactors. In this study, the COD:sulphate radio of 2.4 was not changed in two sequencing batch bioreactors (SBR), however, one SBR was tested at low sulphate concentration (0.4 g $SO_4^{2-}.L^{-1}$) and the second at high sulphate concentration (2.5 g $SO_4^{2-}.L^{-1}$). By means of this test, this study aims to optimize the start-up phase of biological sulphate reduction in SBR.

5.2 Material and methods

5.2.1 Source of biomass

A sequencing batch reactor (SBR) was inoculated with sulphate reducing biomass and used as a control reactor (L). The SBR L was operated with a volatile suspended solid (VSS) concentration of 8.9 ($\pm$ 1.5) g $VSS.L^{-1}$ and total suspended solid (TSS) concentration of 14.1 ($\pm$ 1.7) g $TSS.L^{-1}$. The second SBR, named H, was inoculated with anaerobic sludge from a methanogenic process treating vinasse wastewater. This sludge contained a volatile suspended solid concentration of 36.5 ($\pm$ 0.6) g $VSS.L^{-1}$ and a total suspended solid concentration of 75.6 ($\pm$ 1) g $TSS.L^{-1}$.

5.2.2 Synthetic wastewater

The composition of the synthetic wastewater used in this research was as follows (in $mg.L^{-1}$): NH_4Cl (300), $MgCl_2 \cdot 6H_2O$ (120), KH_2PO_4 (200), KCl (250), $CaCl_2 \cdot 2H_2O$ (15), yeast extract (20) and 0.5 mL of a mixture of micronutrients. The micronutrients solution contained (in $mg.L^{-1}$): $FeCl_2 \cdot 4H_2O$ (1500), $MnCl_2 \cdot 4H_2O$ (100), EDTA (500), H_3BO_3 (62), $ZnCl_2$ (70), $NaMoO_4 \cdot 2H_2O$ (36), $AlCl_3 \cdot 6H_2O$ (40), $NiCl_3 \cdot 6H_2O$ (24), $CoCl_2 \cdot 6H_2O$ (70), $CuCl_2 \cdot 2H_2O$ (20) and HCl 36 % (1 mL) (Villa-Gomez et al., 2012). Sodium lactate and sodium sulphate were used as electron donor and acceptor, respectively. The synthetic wastewater influent pH was adjusted to 6.0 with NaOH (1 M) or HCl (1 M). All reagents used in this study were of analytical grade.

5.2.3 Reactor set up

Two SBR of 6 L active volume were used in this study (Figure 5-1). These SBR were fed and discharged with two peristaltic pumps (Masterflex L/S). The reactors operated 3 cycles of 8 h per day. The schedule of each cycle was: 2 minutes to discharge, two minutes of feeding, 3 hours of agitation stating from time zero and 5 hours of settling from 3 to 8 h of the cycle. 1 L of supernatant was removed and the same volume was fed, resulting in a liquid flow in (Q_{in}) of 3 L. d^{-1} or an HRT = 2 d. The stirring system consisted of an axis with two propels, and the speed was fixed at 120 rpm. The temperature was controlled at 30 °C, with a water bath (Cole Parmer, Polystat 12112-00).

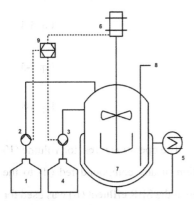

Figure 5-1. Schematic of the sequencing batch reactors
Components: 1) Influent tank, 2) Peristaltic pump, 3) Peristaltic pump, 4) Effluent tank, 5) Heating system (water bath), 6) Stirring system, 7) SBR, 8) Sampling port and 9) timer.

5.2.4 Experimental design

Two different biomasses were tested in two SBR, the control (L) and the experiment (H) bioreactor. The SBR L with sulphate reducing biomass was fed with 417 mg $SO_4^{2-}.L^{-1}$ (140 mg $S\text{-}SO_4^{2-}.L^{-1}$) of sulphate and 1 g $COD.L^{-1}$ of lactate. The SRB H was fed with sulphate at 2502 mg $SO_4^{2-}.L^{-1}$ (840 mg $S\text{-}SO_4^{2-}.L^{-1}$) and 6 g $COD.L^{-1}$ of lactate. In both SBR, the influent $COD:SO_4^{2-}$ ratio was maintained at 2.4. Therefore, in the SRB H the sulphate and lactate concentration were 6 times higher than those used for the SBR L. The SBR L was operated at an HRT of 2 d along the experiment. The SBR H was operated under the following schedule: 1) initially, 8 days at 2 d HRT, 2) followed by 6 days of batch conditions, the influent and effluent pumps were stopped at this time, and 3) the last 20 days, it was operated at 2 d HRT.

5.2.5 Evaluation of the performance of the reactor

The activity in the reactor was evaluated in terms of the loading rates (*LR*) compared to the removal rates (*RR*), fraction (*f*) of a component in the effluent, removal efficiencies (*RE*). However, the robustness of the process was evaluated in terms of resistance, resistance index and resilience according the following equations:

$$LR = \frac{Q(A)}{V}$$
Eq. 5-1

$$RR = \frac{Q(A-B)}{V}$$
Eq. 5-2

$$f = \frac{(A-B)}{A}$$
Eq. 5-3

$$RE = \left(\frac{(A-B)}{A}\right) \times 100 \qquad\qquad\qquad \text{Eq. 5-4}$$

$$Resistance = \frac{RE}{\Delta t} \qquad\qquad\qquad\qquad \text{Eq. 5-5}$$

$$Resistance\ index = \frac{Resistance\ of\ reactor\ experiment}{Resistance\ of\ reactor\ control} \qquad\qquad \text{Eq. 5-6}$$

The flow rate (Q) of 3 L.d^{-1} was equivalent to 3 cycles.d^{-1}. The operational reactor volume (V) of the SBR was expressed in L. The initial concentration of any compound fed (A) to the biomass in the SBR or the concentration of any compound in the SBR effluent (B) was used for calculation, $e.g.$ compounds like the initial SO_4^{2-}, S^{2-} SO_4^{2-}, lactate or total COD (TCOD) concentrations. The time difference (Δt) is defined between the time of starting a new condition (t_{NC}) and the time necessary to reach a removal efficiency $\geq 80\%$ of any compound ($t_{RE \geq 80\%}$).

5.2.6   Chemical and biological analysis

The pH was measured off-line with a sulphide resistant electrode (Prosense, Oosterhout, The Netherlands). Sulphate and ammonia were analyzed by ion chromatography (DIONEX 100) using conductivity detection (Mottet $et\ al.$, 2014). The volatile fatty acids (VFA) concentrations (acetate, propionate,iso-butyrate, butyrate, iso-valerate and valerate) were measured in the soluble phase using a gas chromatograph (GC-800 Fisons Instrument) equipped with a flame ionization detector (Mottet $et\ al.$, 2014). Lactate was analyzed by HPLC as reported in the literature (Quéméneur $et\ al.$, 2011). The VSS, TSS and apart the sulphide or total dissolved sulfide (TDS, by the methylene blue method) were measured according to the procedures outlined in Standard Methods (1992).

5.3   Results

5.3.1   Anaerobic sulphate reduction in a SBR at low sulphate concentrations (L)

Figure 5-2A-D describes the performance of the SBR L, the flow rates are shown to depict the mass flown in and out of the system in a daily basis. When the S-SO_4^{2-}-RR reached the value equal to the S-SO_4^{2-}-LR, the sulphate RE was assumed 100%. The time to reach $> 80\%$ of sulphate RE ($t_{RE \geq 80\%}$) was after 12 d of SBR operation. During the first 12 d of operation, 22 (± 15)% sulphate RE was observed. After (12 d) this time, the SBR L performance was considered steady state until the end of the operation (34 d), the sulphate RE was 90 (± 9)% (Figure 5-2A). Sulphide concentrations were as low as 27 (± 21) and 60 (± 25) mg S^{2-}.L^{-1}, respectively, during and after the first 12 d of operation. Therefore, the sulphide production rates ranged 13-30 mg $S^{2-}.L^{-1}d^{-1}$, respectively (Figure 5-2A).

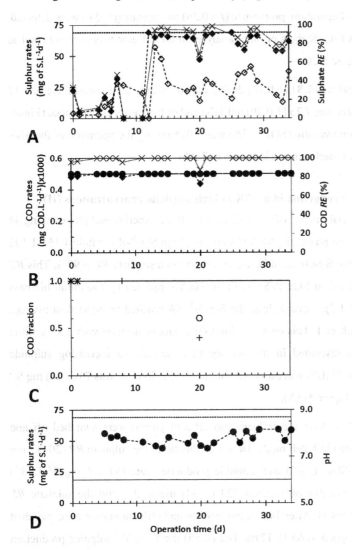

Figure 5-2. Performance of an SBR at low influent sulphate concentrations (SBR L)
Profiles of A) $S-SO_4^{2-}$ loading rate (⋯), $S-SO_4^{2-}$ removal rate (◆), production rate of $S-S^{2-}$ (◇) and sulphate removal efficiency (x); B) the COD loading rate (⋯), COD removal rate (◆), the lactate removal rate (●) and COD removal efficiency (x); C) the fraction composition of the effluent COD: lactate (Δ), acetate (○), propionate (+) and iso-butirate (*); D) the pH evolution along the process (●), the $S-SO_4^{2-}$ loading rate (⋯) and the pH of the influent at 6.0 (—). The SBR L was operated at constant 8 h.cycle^{-1}.

The lactate *RR* or consumption rate was the same as 100% *RE* along the SBR operation (34 d). Likewise, the TCOD *RE* was 99 (± 3)% along the 34 d of operation (Figure 5-2B). The effluent

COD composition was analysed during the time of SBR L operation (Figure 5-2C). During the first two days, iso-butyrate (f = 1) was detected. Later, propionate (f = 1) was detected as sole VFA at time 8. At time 20 of operation, propionate (f = 0.39) and acetate (f = 0.6) were detected. The presence of COD as VFA, alcohols or other analytes in the effluent was very punctual as described above and in Figure 5-2C.

Furthermore, the effluent pH was 7.81 ($\pm$ 0.14) during the non-steady performance, the first 12 d of SRB operation, and later was 7.75 ($\pm$ 0.29) until the end of the steady performance (Figure 5-2D). The substrate to biomass ratio (SO_4^{2-}:VSS) was 0.025 during the operation of the SBR L. A summary on the SBR L performance is shown in Table 5-1.

5.3.2 Anaerobic sulphate reduction in a SBR at high sulphate concentrations (H)

Figure 5-3A-D shows the performance of the SBR H in the three experimental phases along 34 d. In the beginning of the first phase, the SBR H showed a high SO_4^{2-}-RR, within 1,165-1,192 mg $SO_4^{2-}.L^{-1}d^{-1}$ (392-401 mg $S-SO_4^{2-}.L^{-1}d^{-1}$), corresponding to a sulphate RE > 90%. This RE could be supported by the initial SO_4^{2-}:VSS ratio of 0.068 supported by the initial biomass concentration (36.5 g $VSS.L^{-1}$). Nevertheless, the $S-SO_4^{2-}$-RR reached an equivalent average RE of 62 ($\pm$ 25)% during phase I. This could be due to washout of bacteria with poor settling velocity (> 5 h) that are discarded in the settling time period. An increasing sulphide concentration (148 $\pm$ 97 mg $S^{2-}.L^{-1}$) was observed and the production rate was 74 ($\pm$ 48) mg $S^{2-}.L^{-1}d^{-1}$ during this phase I (Figure 5-3A).

During phase II (from 8.3-14.3 d), the influent and effluent pumps were switched off and therefore the SBR H operated in batch mode. Hence, at this time, the sulphate RR (20$\pm$17 mg $SO_4^{2-}.L^{-1}d^{-1}$ or 7$\pm$6 mg $S-SO_4^{2-}.L^{-1}d^{-1}$) and sulphide production rate (54 $\pm$ 6 mg $S^{2-}.L^{-1}d^{-1}$) decreased, the sulphide concentration reached 324 ($\pm$ 34) mg $S^{2-}.L^{-1}$ and the sulphate RE reached 95 ($\pm$ 4)%, in the SBR H. After 14.3 d, the influent and effluent pumps were switched on and the sulphate RE dropped to 65 ($\pm$ 12)%, 142 ($\pm$ 55) mg $S^{2-}.L^{-1}d^{-1}$ sulphide production rate was observed corresponding to a sulphide concentration of 284 ($\pm$ 111)% mg $S^{2-}.L^{-1}$, at the beginning of phase III.

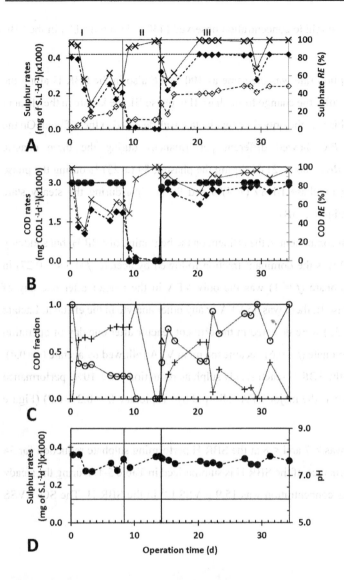

Figure 5-3. Performance of an SBR at high influent sulphate concentrations (SBR H)
Profiles of A) S-SO$_4$$^{2-}$ loading rate (⋯), S-SO$_4$$^{2-}$ removal rate (◆), production rate of S-S^{2-} (◇) and sulphate removal efficiency (×); B) COD loading rate (⋯), COD removal rate (◆), lactate removal rate (●) and COD removal efficiency (×); C) the fraction composition of the effluent COD: lactate (Δ), acetate (○), propionate (+) and iso-butirate (*); D) the pH evolution along the process (●), the S-SO$_4$$^{2-}$ loading rate (⋯) and the pH of the influent at 6.0 (—). The SBR H was operated in three periods (I: 8 h.cycle^{-1} SBR; II: batch for 6 days; and III: 8 h.cycle^{-1} SBR)

However, from day 20 to 34 of operation (phase III) the performance reached a steady state sulphate RE that averaged 96 ($\pm$ 10)% corresponding to a sulphate RR of 1193 ($\pm$140) mg SO$_4$$^{2-}$ .L^{-1}d^{-1} (or 401$\pm$47 mg S-SO$_4$$^{2-}$.L^{-1}d^{-1}). The sulphide production rate was 201 ($\pm$ 46) mg S^{2-}.L^{-}

^{1}d^{-1} for the highest average sulphide concentration observed (440 ± 41 mg S^{2-}.L^{-1}) in the SBR H (Figure 5-3A).

The lactate *RR* or consumption rate was the same as 100 % *RE* also in the SBR H along the operation time (34 d). Only after the change from phase II to phase III, the lactate in the effluent corresponded to a f = 0.6, this was the only time that lactate was detected in the effluent. On the other hand, the TCOD *RR* showed different performances during the three phases, corresponding to a TCOD *RE* of 60 (± 21)% during the phase I, 84 (± 23) % during the phase II, while at the beginning of phase III 89 (± 4)% and 90 (± 7)% during the steady state performance of phase III (Figure 5-3B).

Acetate (f = 1) was the main component in the effluent on the beginning of SBR H, but suddenly propionate (f = 0.6 ± 0.27) was the dominant fraction followed by acetate (f = 0.4 ± 0.27) in phase I. At phase II, propionate (f = 1) was the only VFA in the reactor after one day of operation. At the end of phase II, there was no VFA or any other analyte in the effluent. Lactate (f = 0.6) and acetate (f = 0.39) were detected in the effluent reactor after one day of operation of phase III. At day 20 propionate (f = 0.6) became the main VFA followed by acetate (f = 0.4). After 20 days of operation the SBR H reach steady sulphate reduction (96 ± 10%) performance and acetate (f = 0.8 ± 0.2) was the major VFA followed by propionate (f = 0.2 ± 0.2) (Figure 5-3C).

The effluent pH observed was > 7 and < 8 at the SBR H performing sulphate reduction for 34 d (Figure 5-3D). The performance of the SBR H is summarized in Table 5-1. During the steady state performance, the VSS concentration was 15.9 g VSS.L^{-1} in the SBR H. The SO$_4{}^{2-}$:VSS ratio was equal to 0.15.

Table 5-1. Operational condition and performance of the control (L) and experiment (H) SBR

Phase	OT (d)	OM	Stage (d)	COD-LR	SO$_4^{2-}$-LR	S-SO$_4^{2-}$-LR	L-RR	L-RE (%)	TCOD-RR	TCOD-RE (%)	SO$_4^{2-}$-RR	SO$_4^{2-}$-RE (%)	S-SO$_4^{2-}$-RR	S^{2-}-PR	S^{2-} conc. (mg.L^{-1})	pH
			SBR L													
I	34	SBR	NSP (12)	500	209	69	500	100	495±14	99±3	45±33	22±15	15±11	13±10	27±21	7.81±0.14
			SP (22)								189±19	90±9	63±6	30±12	60±25	7.75±0.29
			SBR H													
I	8	SBR	NSP (8)	3,000	1,250	420	3,000	100	1,793±616	60±21	737±335	62±25	249±112	74±48	148±97	7.51±0.32
II	6	Batch	(6)	0	0	0	0	100	155±234	84±23	20±17	95±4	7±6	54±6	324±34	7.65±0.27
III	20	SBR	NSP (6)	3,000	1,250	420	2,943±99	98±3	2,671±106	89±4	785±156	65±12	265±52	142±55	284±111	7.62±0.11
			SP (14)	3,000	1,250	420	3,000	100	2,688±227	90±7	1,193±140	96±10	401±47	201±46	440±41	7.57±0.13

Note: Loading (LR), removal (RR) and production (PR) rates are shown with the units of mg.L^{-1}.d^{-1}. OT = Operation time, OM= Operation mode, NSP= non steady performance, SP= steady performance

Table 5-2. Stoichiometric reactions in the sequencing batch bioreactors

Stoichiometric reaction	ΔG$^{0'}$ (kJ.mol^{-1})	
3 Lactate$^-$ → 2 Propionate$^-$ + Acetate$^-$ + CO$_2$	-54.9	Eq. 5-7
2 Lactate$^-$ + SO$_4^{2-}$ → 2 Acetate$^-$ + HS$^-$ + 2HCO$_3^-$ + H$^+$	-160.1	Eq. 5-8
2Lactate$^-$ + 3SO$_4^{2-}$ → 6 HCO$_3^-$ + HS$^-$ + H$^+$	-255.3	Eq. 5-9
Propionate$^-$ + 0.75 SO$_4^{2-}$ → Acetate$^-$ + HCO$_3^-$ + 0.75 HS$^-$ + 0.25 H$^+$	-37.7	Eq. 5-10
Propionate$^-$ + 1.75 SO$_4^{2-}$ → 3 HCO$_3^-$ + 1.75 HS$^-$ + 0.25 H$^+$	-85.4	Eq. 5-11
Acetate$^-$ + SO$_4^{2-}$ → 2 HCO$_3^-$ + HS$^-$	-48	Eq. 5-12
4 H$_2$ + SO$_4^{2-}$ + H$^+$ → HS$^-$ + 4 H$_2$O	-151.9	Eq. 5-13
4 H$_2$ + HCO$_3^-$ + H$^+$ → CH$_4$ + 3 H$_2$O	-135.6	Eq. 5-14
Acetate$^-$ + H$_2$O → CH$_4$ + HCO$_3^-$	-31.1	Eq. 5-15

Table 5-3. Comparison and resistance of sulphate reduction in SBR

SO_4^{2-} RE (%)	COD RE (%)	$t_{RE \geq 80\%}$ (d)	COD:SO_4^{2-} ratio	HRT (d)	Source of biomass	Electron donor	Reactor	Resistance (using Eq. 5-5, dimensionless)	Resistance index (using Eq. 5-6, dimensionless)	References
>80	55 - 65	325	2.39	1	Anaerobic granular sludge	Lactate	UASB	0.25 (80/325)		Bertolino et al. (2012)
>80		~ 85	2	4	Anaerobic granular sludge	Ethanol	UASB	0.95 (80/85)		Velasco et al. (2008)
99		15	2.6	>3 d (48 h. cycle⁻¹)	Anaerobic sludge	Butanol	packed SBR	5.3 (80/15)		Sarti and Zaiat (2011)
98	<45	15	2	0.4	Shuangcheng moat sediment (Heilongjiang, China)	Lactate	CSTR	5.3 (80/15)	1.6 (8.6/5.3) and 0.75 (4/5.3)	Zhao et al. (2008)
90±9	99±3	12	2.4	2	Sulphate reducing biomass	Lactate	SBR L (control reactor)	6.6 (80/12)		In this study
95±4	95±23	9.3	2.4	2	Anaerobic sludge	Lactate	SBR H	8.6 (80/9.3)	1.3 (8.6/6.6)	In this study
96±10	90±7	20	2.4	2	Anaerobic sludge	Lactate	SBR H	4 (80/20)	0.6 (4/6.6)	In this study
		1 (from 8.3 to 9.3)	2.4		Anaerobic sludge	Lactate	SBR H	80 (80/1)	12 (80/6.6)	In this study
		6 (from 14.3 to 20)	2.4		Anaerobic sludge	Lactate	SBR H	13.6 (80/6)	2 (13.6/6.6)	In this study

5.4 Discussion

5.4.1 Sulphate reduction process in the SBR

This study showed that that high influent sulphate concentrations promoted the higher removal of sulphate and reduce the time of start-up in the SBR. The specific growth rate of a pure culture of SRB is promoted by high sulphate concentrations, 2.5 g $SO_4^{2-}.L^{-1}$ (Al-Zuhair et al., 2008). There is evidence that SRB growth is promoted by high sulphate concentrations also when perform in an anaerobic sludge. In this study, the control SBR (L) showed a poor sulphate RE ($22 \pm 15\%$) during a lag phase of 12 d and in comparison to the RE showed by the SBR H (62 $\pm 25\%$) exposed to high initial sulphate concentrations (2.5 g $SO_4^{2-}.L^{-1}$) (Figure 5-2A and Figure 5-3A), both SBR used a COD:SO_4^{2-} ratio of 2.4. Using this ratio, a sulphate RE > 90 % is guaranteed (Torner-Morales and Buitrón, 2010). Moreover, a COD:SO_4^{2-} ratio < 2 could be optimal to outcompete the methanogenes (O'Reilly and Colleran, 2006). Additionally, the high diversity of bacteria type in the anaerobic sludge can be another reason for the sulphate removal observed using anaerobic sludge in the SBR H, when compared to an already enriched sulphate reducing biomass (Guo et al., 2014).

In both SBR, L and H, the lactate was used efficiently (100 %) from the starting point of the experiment till the end. The SBR H showed low TCOD removal efficiency ($60 \pm 21\%$ and a rate of $1,793 \pm 616$ mg COD.$L^{-1}d^{-1}$) when compared to the control reactor (removal efficiency of $99 \pm 3\%$ and 495 ± 14 mg COD.$L^{-1}d^{-1}$). The propionate fraction became dominant in the effluent, compared to the acetate fraction, in the first phase of SRB H operation. Such evolution suggests that hydrolytic-fermentative bacteria outcompete the SRB for the fermentation of lactate to propionate, this is supported by the specific growth rates of hydrolytic-fermentative bacteria ($\mu_{VFA\ mixture} \gg 1.2$ d^{-1}) (Escudié et al., 2005) in comparison to SRB (*Desulfovibrio desulfuricans*, $\mu_{Lactate} \sim 0.052$ d^{-1}) (Cooney et al., 1996) and to methanogenic archaea ($\mu_{Acetate} \sim$ 0.15-0.55 d^{-1}) (Vincent O'Flaherty et al., 1998). Furthermore, the conversion of lactate to propionate is also supported by the free energy of formation ($\Delta G^{0\prime}$, Table 5-2) given by Eq. 5-7. A minor fraction of lactate might be used by SRB as described in Eq. 5-8, however, this is not the case of Eq. 5-9 that takes place when there is no competition for lactate by other microorganism. The $\Delta G^{0\prime}$ given by Eq. 5-10 suggests that sulphate reduction is possible using propionate as electron donor rather than the use of lactate. During phase I of SBR H, the propionate was the dominant fraction in the effluent and the sulphate RE was 62 (± 25)% indicating a developing population of propionate consuming SRB.

In phase II, bacteria consumed the remaining propionate and the sulphate *RE* improved to 95 (± 4) % in SRB H. Similar improvements on the removal of sulphate have been reported when there is a change on the operation mode of the reactor, *e.g.* from UASB to CSTR with biomass recirculation (Boshoff *et al.*, 2004). This suggests that a change on the operating conditions of the reactor, a modification like increasing the HRT, is beneficial for sulphate reduction. Increasing the HRT optimize the contact time of bacteria with substrates, hence, slow growing SRB (in comparison to hydrolytic-fermentative bacteria) has the time to produce more propionate oxidizing enzymes or to consume the available propionate with the existing concentration enzyme and to overcome the accumulation (as observed with propionate during the beginning of phase II) or the shock load if is the case.

At phase II in SBR H, the accumulation of sulphide (324 ± 34 mg $S^{2-}.L^{-1}$) could influenced the SRB population growth since SRB is more tolerant to high sulphide concentrations than other bacteria and archaea. Continuous bioreactors, like inverse fluidized bed bioreactors, with biomass recirculation can operate at a sulphide concertation of 1,200 mg.L^{-1} without affecting the COD and sulphate removal efficiency (Celis-García *et al.*, 2007). Additionally, the toxicity of sulphide concentration is higher when reactors are operated at long HRT (Kaksonen *et al.*, 2004).

Decreasing the HRT might hamper the performance of SBR. After a change on the HRT, the sulphate *RE* was 65 (± 12)% and the TCOD *RE* 89 (± 4)% in beginning of the third phase. Non steady sulphate removal efficiency was observed for 6 d (from 14 to 20 d) of operation. After 20 days of SBR H operation, sulphate removal efficiency at steady conitions was achieved (96 ± 10%), acetate was the major by-product and propionate became the minor component in the TCOD remaining in the effluent (~ 10% of the COD fed).

Propionate removal is linked to the removal of sulphate due to the activity of SRB growing on the oxidation of this VFA. The lack of propionate consuming SRB might hamper the performance of bioreactors (O'Flaherty and Colleran, 1999). Furthermore, propionate, hydrogen and carbon dioxide are better utilized by SRB in the presence of sulphate (Lens *et al.*, 1998; O'Flaherty *et al.*, 1998; Qatibi *et al.*, 1990).

5.4.2   Robustness of biological sulphate reduction in SBR

According Cabrol *et al.* (2012), Eq. 5-5 should quantitatively describe the situation of a reactor when a change has been made to an operational variable (HRT, influent concentration of a random compound fed, etc.). Hereby, the resistance equation (Eq. 5-5) is used to describe and compare the lag phase during the start-up of the SBR to other reactor performances (Table 5-3).

The ratios obtained using Eq. 5-5 expresses the resistance of the biomass to remove > 80% of sulphate, a small value expresses high resistance. The resistance can be a consequence of the process variables used during the bioreactor operation. Table 5-3 shows that sulphate reduction take place at > 80% in different reactor configurations, but all of them showed different levels of resistance. For instance, the experiments of Bertolino et $al.$ (2012) and Velasco et $al.$ (2008) showed a resistance < 1, the reason for this might be the lower COD:SO_4^{2-} ratios used previous to the obtained sulphate RE > 80%. On the other hand, resistance values of 5.3 could be observed in the research of Sarti and Zaiat (2011) and Zhao et $al.$ (2008) where the COD:SO_4^{2-} ratios used were ≥ 2. Additionally, Sarti and Zaiat (2011) used butanol as electron donor and long cycles of 2 d (HRT > 3 d) in the SBR that could reduce limiting conditions due to hydrodynamics. Zhao et $al.$ (2008), used a short HRT (0.4 d) in a CSTR but used a sea sediment as inoculum. In this study, the resisance was 6.6 for SBR L at influent sulphate concentration of 417 mg SO_4^{2-}.L^{-1} (140 mg S-SO_4^{2-}.L^{-1}), this indicate slightly lower resistance compared those shown in the experiments by Sarti and Zaiat (2011) and Zhao et $al.$ (2008).

The SBR H, evaluated (using the Eq. 5-5) at the time 9.3, sowed a resistance of 8.6. Nevertheless, the resistance was 80 for the SBR H evaluated at 1 d, sulphate RE was > 80% one day after stopping the influent and effluent pumps, from phase I to II. If the SBR H is evaluated at 20 d, the resistance of sulphate reduction is slightly larger compared to Sarti and Zaiat (2011) and Zhao et $al.$ (2008). However, the resistance is the lowest (13.6) considering 6 days after the beginning of phase III (Table 5-3).

Moreover, according to Cabrol et $al.$ (2012), the Eq. 5-6 can used to compare the performance of the reactor experiment (SBR H) to a reactor used as control (SBR L) and the process performance can be quantified during a disturbance. Then, a ratio < 1 from Eq. 5-6 indicates that the resistance was lower in the control reactor and a ratio > 1 indicates that the reactor control improved the performance on the conditions tested in comparison to the control reactor.

5.5   Conclusions

This research showed that the start-up phase of biological sulphate removal from synthetic wastewater can be optimized or shortened at high influent sulphate concentrations (2.5 g SO_4^{2-} .L^{-1}). Propionate was the major VFA observed in the effluent during non-steady state sulphate reduction, while in steady performance, acetate was the major end product. Therefore, lactate was fermented to propionate and the last was further used as electron donor by SRB to remove sulphate and produce acetate at steady performance. These results indicated that the propionate degrading SRB play a major role in the robustness of sulphate reduction, during the start-up phase of SBR. The resistance of sulphate removal was inferior in SBR H compared to the reactor control SBR L (exposed to ideal conditions for sulphate removal) and to other sulphate removal processes reported in the literature.

5.6 References

Al-Zuhair, S., El-Naas, M.H., Al-Hassani, H., 2008. Sulfate inhibition effect on sulfate reducing bacteria. J. Biochem. Technol. 1, 39–44.

APHA, 1999. Standard Methods for the Examination of Water and Wastewater, 20th ed. Washington, USA.

Bertolino, S.M., Rodrigues, I.C.B., Guerra-Sá, R., Aquino, S.F., Leão, V.A., 2012. Implications of volatile fatty acid profile on the metabolic pathway during continuous sulfate reduction. J. Environ. Manage. 103, 15–23. doi:10.1016/j.jenvman.2012.02.022

Boshoff, G., Duncan, J., Rose, P., 2004. Tannery effluent as a carbon source for biological sulphate reduction. Water Res. 38, 2651–2658. doi:10.1016/j.watres.2004.03.030

Cabrol, L., Malhautier, L., Poly, F., Roux, X. Le, Lepeuple, A.S., Fanlo, J.L., 2012. Resistance and resilience of removal efficiency and bacterial community structure of gas biofilters exposed to repeated shock loads. Bioresour. Technol. 123, 548–557. doi:10.1016/j.biortech.2012.07.033

Celis-García, L.B., Razo-Flores, E., Monroy, O., 2007. Performance of a down-flow fluidized bed reactor under sulfate reduction conditions using volatile fatty acids as electron donors. Biotechnol. Bioeng. 97, 771–779. doi:10.1002/bit.21288

Cooney, M.J., Roschi, E., Marison, I.W., Comninellis, C., von Stockar, U., 1996. Physiologic studies with the sulfate-reducing bacterium *Desulfovibrio desulfuricans*: evaluation for use in a biofuel cell. Enzyme Microb. Technol. 18, 358–65.

Dries, J., De Smul, A., Goethals, L., Grootaerd, H., Verstraete, W., 1998. High rate biological treatment of sulfate-rich wastewater in an acetate-fed EGSB reactor. Biodegradation 9, 103–111. doi:10.1023/A:1008334219332

Escudié, R., Conte, T., Steyer, J.P., Delgenès, J.P., 2005. Hydrodynamic and biokinetic models of an anaerobic fixed-bed reactor. Process Biochem. 40, 2311–2323. doi:10.1016/j.procbio.2004.09.004

Guo, X., Wang, C., Sun, F., Zhu, W., Wu, W., 2014. A comparison of microbial characteristics between the thermophilic and mesophilic anaerobic digesters exposed to elevated food waste loadings. Bioresour. Technol. 152, 420–428. doi:10.1016/j.biortech.2013.11.012

Kaksonen, A.H., Franzmann, P.D., Puhakka, J.A., 2004. Effects of hydraulic retention time and sulfide toxicity on ethanol and acetate oxidation in sulfate-reducing metal-precipitating fluidized-bed reactor. Biotechnol. Bioeng. 86, 332–343. doi:10.1002/bit.20061

Lens, P.N., Dijkema, C., Stams, A.J., 1998. 13C-NMR study of propionate metabolism by sludges from bioreactors treating sulfate and sulfide rich wastewater. Biodegradation 9, 179–186. doi:10.1023/A:1008395724938

Macarie, H., Guyot, J.P., 1995. Use of ferrous sulphate to reduce the redox potential and allow the start-up of UASB reactors treating slowly biodegradable compounds: application to a wastewater containing 4-methylbenzoic acid. Environ. Technol. 16, 1185–1192. doi:10.1080/09593331608616354

Mapanda, F., Nyamadzawo, G., Nyamangara, J., Wuta, M., 2007. Effects of discharging acid-mine drainage into evaporation ponds lined with clay on chemical quality of the surrounding soil and water. Phys. Chem. Earth, Parts A/B/C 32, 1366–1375. doi:10.1016/j.pce.2007.07.041

Mottet, A., Habouzit, F., Steyer, J.P., 2014. Anaerobic digestion of marine microalgae in different salinity levels. Bioresour. Technol. 158, 300–6. doi:10.1016/j.biortech.2014.02.055

O'Flaherty, V., Colleran, E., 1999. Effect of sulphate addition on volatile fatty acid and ethanol degradation in an anaerobic hybrid reactor. I: process disturbance and remediation. Bioresour. Technol. 68, 101–107. doi:10.1016/S0960-8524(98)00145-X

O'Flaherty, V., Lens, P., Leahy, B., Colleran, E., 1998. Long-term competition between sulphate-reducing and methane-producing bacteria during full-scale anaerobic treatment of citric acid production wastewater. Water Res. 32, 815–825. doi:10.1016/S0043-1354(97)00270-4

O'Flaherty, V., Mahony, T., O'Kennedy, R., Colleran, E., 1998. Effect of pH on growth kinetics and sulphide toxicity thresholds of a range of methanogenic, syntrophic and sulphate-reducing bacteria. Process Biochem. 33, 555–569. doi:10.1016/S0032-9592(98)00018-1

O'Reilly, C., Colleran, E., 2006. Effect of influent COD/SO_4^{2-} ratios on mesophilic anaerobic reactor biomass populations: physico-chemical and microbiological properties. FEMS Microbiol. Ecol. 56, 141–153. doi:10.1111/j.1574-6941.2006.00066.x

Qatibi, A.I., Bories, A., Garcia, J.L., 1990. Effects of sulfate on lactate and C2-, C3- volatile fatty acid anaerobic degradation by a mixed microbial culture. Antonie van Leeuwenhoek 58, 241–8.

Quéméneur, M., Hamelin, J., Latrille, E., Steyer, J.-P., Trably, E., 2011. Functional versus phylogenetic fingerprint analyses for monitoring hydrogen-producing bacterial populations in dark fermentation cultures. Int. J. Hydrogen Energy 36, 3870–3879. doi:10.1016/j.ijhydene.2010.12.100

Sarti, A., Zaiat, M., 2011. Anaerobic treatment of sulfate-rich wastewater in an anaerobic sequential batch reactor (AnSBR) using butanol as the carbon source. J. Environ. Manage. 92, 1537–1541. doi:10.1016/j.jenvman.2011.01.009

Sipma, J., Osuna, M.B., Lettinga, G., Stams, A.J.M., Lens, P.N.L., 2007. Effect of hydraulic retention time on sulfate reduction in a carbon monoxide fed thermophilic gas lift reactor. Water Res. 41, 1995–2003. doi:10.1016/j.watres.2007.01.030

Torner-Morales, F.J., Buitrón, G., 2010. Kinetic characterization and modeling simplification of an anaerobic sulfate reducing batch process. J. Chem. Technol. Biotechnol. 85, 453–459. doi:10.1002/jctb.2310

Velasco, A., Ramírez, M., Volke-Sepúlveda, T., González-Sánchez, A., Revah, S., 2008. Evaluation of feed COD/sulfate ratio as a control criterion for the biological hydrogen sulfide production and lead precipitation. J. Hazard. Mater. 151, 407–413. doi:10.1016/j.jhazmat.2007.06.004

Villa-Gomez, D.K., Papirio, S., van Hullebusch, E.D., Farges, F., Nikitenko, S., Kramer, H., Lens, P.N.L., 2012. Influence of sulfide concentration and macronutrients on the characteristics of metal precipitates relevant to metal recovery in bioreactors. Bioresour. Technol. 110, 26–34. doi:10.1016/j.biortech.2012.01.041

Zandvoort, M.H., Geerts, R., Lettinga, G., Lens, P.N.L., 2003. Methanol degradation in granular sludge reactors at sub-optimal metal concentrations: role of iron, nickel and cobalt. Enzyme Microb. Technol. 33, 190–198. doi:10.1016/S0141-0229(03)00114-5

Zandvoort, M.H., van Hullebusch, E.D., Gieteling, J., Lettinga, G., Lens, P.N.L., 2005. Effect of sulfur source on the performance and metal retention of methanol-fed UASB reactors. Biotechnol. Prog. 21, 839–850. doi:10.1021/bp0500462

Zhao, Y., Ren, N., Wang, A., 2008. Contributions of fermentative acidogenic bacteria and sulfate-reducing bacteria to lactate degradation and sulfate reduction. Chemosphere 72, 233–242. doi:10.1016/j.chemosphere.2008.01.046

Chapter 6

The effect of nitrogen and electron donor feast-famine conditions on biological sulphate reduction in inorganic wastewater treatment

Abstract

Transient feeding conditions might hamper biological sulphate reduction (BSR) during bioreactor operation. This research studied the effect of NH_4^+ and electron donor feast to famine conditions on BSR in batch bioreactors (agitated at 120 rpm and 30 °C). Lactate (COD) was fixed at 1000 mg.L^{-1} along the experiments. The NH_4^+ was either included (300 mg. L^{-1}) or completely excluded from the synthetic wastewater. The electron feast to famine conditions were stablished by modification of the initial sulphate concentration (417, 666 and 1491 mg $SO_4^{2-}.L^{-1}$). The sulphate removal efficiency was > 95% using electron donor feast conditions and decreased till < 50% in the electron donor famine conditions. Using NH_4^+ feast conditions, the first order kinetic constant (k_1) of sulphate removal decreased 4 % compared to the NH_4^+ famine conditions. The specific electron donor utilization rate (4.39 mg TCOD. mg VSS^{-1} d^{-1}, r^2=0.9895) improved 16.6 % using NH_4^+ feast conditions, in comparison to NH_4^+ famine conditions (3.66 mg TCOD. mg VSS^{-1} d^{-1}, r^2=0.99) during the sulphate removal. The electron donor flow to sulphate reduction increased simultaneously to electron donor famine conditions, this research showed that high sulphate removal efficiencies (> 90%) are not possible at COD famine conditions.

Keywords: Sulphate removal, sulphidogenesis, feast-famine conditions, transient feeding conditions

6.1 Introduction

Inorganic wastewaters rich in sulphate need to be treated and not discharged untreated into nature, otherwise, the production of toxic, corrosive and poisonous sulphide gas can take place. Biological treatment of sulphate rich inorganic wastewaters is a process wherein the removal of sulphate is carried out under controlled conditions and further recovery of sulphide is possible (Lens *et al.*, 2002). The sulphate reducing bacteria (SRB) reduce the sulphate into sulphide. Later, sulphide might be used for the recovery of heavy metals from the same wastewater, like in acid mine drainage (Lewis, 2010).

SRB are capable to use many sources of carbon as electron donors (Liamleam and Annachhatre, 2007) and are capable to perform autotrophic or heterotrophic sulphate reduction (Plugge *et al.*, 2011) but lack hydrolytic enzyme systems. Inorganic wastewaters rich in sulphate lack COD and, therefore, many expensive pure chemicals are used as electron donors during the treatment. Hence, the metabolic flexibility of SRB promises to lower the cost of the treatment of sulphate rich inorganic wastewaters. This might be possible using: hydrogenotrophic pathways (autotrophic metabolism) or lignocellulose as slow release electron donors.

Biological sulphate reduction (BSR) has been applied in different bioreactor configurations (Kaksonen and Puhakka, 2007; Papirio *et al.*, 2013). The selection of the reactor depends on the sulphate content of the inorganic wastewater and the volume to be treated, apart from the requirements to enable the recovery of resources from wastewater. Nevertheless, the proper performance of biological sulphate reducing bioreactors depends on the control of the process variables, *e.g.* electron donor concentration, hydraulic retention time, sulphide concentration, pH and metal concentration in the influent (Kaksonen and Puhakka, 2007).

Transient feeding, feast to famine, conditions as variations on the electron donor or acceptor (sulphate), influent pH can hamper the sulphate reduction in long term continuous bioreactor operation. Notwithstanding, little is known about the consequences of such varying operating conditions on the process dynamics and microbial ecology of sulphate reducing bioreactors. For instance, during sulphate reduction and increasing NH_4^+ famine conditions, *Desulfovibrio desulfuricans* decreased the electron donor uptake, the cell size was negatively affected and the cell carbon content decreased (Okabe *et al.*, 1992). In sulphate reduction experiments under sulphate feast conditions with *Archaeoglobus fulgidus* Strain Z, the biomass formation on sulphate consumed was higher compared to sulphate famine conditions (Habicht *et al.*, 2005). Also, carbon uptake for biomass formation was greater at sulphate feast conditions. Most likely, cells consume electron donors and will choose for the thermodynamically optimal pathway, *e.g.* the shift of biochemical pathways due to sulphate limiting and non-limiting conditions can occur (Habicht *et al.*, 2005). The aim of this research is to study the influence of the NH_4^+ and electron donor feast and famine conditions on the biological sulphate removal of a mesophilic (30 °C) sulphate reducing sludge in batch bioreactors.

6.2   Material and methods

6.2.1   Synthetic wastewater

The composition of the synthetic wastewater used in this research was (in $mg.L^{-1}$): NH_4Cl (300), $MgCl_2 \cdot 6H_2O$ (120), KH_2PO_4 (200), KCl (250), $CaCl_2 \cdot 2H_2O$ (15), yeast extract (20) and 0.5 mL of a mixture of micronutrients. The micronutrients solution contained (in $mg.L^{-1}$): $FeCl_2 \cdot 4H_2O$ (1500), $MnCl_2 \cdot 4H_2O$ (100), EDTA (500), H_3BO_3 (62), $ZnCl_2$ (70), $NaMoO_4 \cdot 2H_2O$ (36), $AlCl_3 \cdot 6H_2O$ (40), $NiCl_3 \cdot 6H_2O$ (24), $CoCl_2 \cdot 6H_2O$ (70), $CuCl_2 \cdot 2H_2O$ (20) and HCl 36 % (1 mL) (Villa-Gomez *et al.*, 2012). Sodium lactate and sodium sulphate were used as electron donor and acceptor, respectively. The source of nitrogen (NH_4Cl) was excluded from the preparation of the synthetic wastewater when it was required by the experiment. The synthetic wastewater influent pH was adjusted to 7.0 with NaOH (1 M) or HCl (1 M). All reagents used in this study were of analytical grade.

6.2.2   Inoculum and batch bioreactor preparation

Sludge was sampled from a sequencing batch reactor performing biological sulphate removal at steady state condition. This sulphate reducing inoculum contained 8.9 (±1.5) g $VSS.L^{-1}$ and was used to seed the bench scale batch reactors. Immediately after sampling, 0.040 L of sulphate reducing inoculum was placed in the batch bioreactors (serum bottles of 0.12 L). The batch bottles were capped with butyl rubber stoppers and sealed whit aluminium caps. After two hour of settling, 0.030 L of supernatant was removed carefully with a syringe. This was followed by addition of new synthetic wastewater with the desired electron donors (lactate) and acceptor (sulphate) concentrations. This procedure was essential to keep the initial amount of inoculum constant. The batch bioreactors were incubated at 30 °C and were agitated at 120 rpm on an orbital shaker (New Brunswick Scientific Innova 2100 platform shaker, Eppendorf, USA).

Table 6-1. Description of the feast and famine conditions for sulphate reduction process in batch experiments

Batch bioreactor	CODLactate (mg COD.L^{-1})	Sulphate (mg SO$_4^{2-}$.L^{-1})	COD:SO$_4^{2-}$ ratio	NH$_4^+$ (mg.L^{-1})
A	1,000	417	2.4	0
B	1,000	666	1.5	0
C	1,000	1,491	0.67	0
A*	1,000	417	2.4	300
B*	1,000	666	1.5	300
C*	1,000	1,491	0.67	300

6.2.3 Experimental design

The sulphate reducing activity of the sludge was investigated under feast and famine conditions. The feast and famine conditions were induced by altering the NH$_4^+$ and sulphate initial concentrations, while the lactate concentration (1,000 mg COD.L^{-1}) was always kept fixed in batch bioreactors. Surplus sulphate was fed in order to reach three different initial concentrations, 417, 666 and 1491 mg SO$_4^{2-}$.L^{-1}, in the respective batch bioreactor. The NH$_4^+$ initial concentration was 0 mg NH$_4^+$.L^{-1} (zero) for the experiments A, B, and C and 300 mg NH$_4^+$.L^{-1} for A*, B* and C* experiments. Table 6-1 overviews the 6 different initial feast and famine conditions for the evaluation of the sulphate reducing activity. All the experiments were performed in triplicate.

6.2.4 Chemical analysis

The volatile suspended solids (VSS), total suspended solids (TSS) and sulphide (total dissolved sulphide, TDS) by the methylene blue reaction were measured according to the procedure outlined in Standard Methods (1992). The pH was measured off-line with a sulphide resistant electrode (Prosense, Oosterhout, The Netherlands). Sulphate and ammonia were analysed by ion chromatography (DIONEX 100) using conductivity detection (Mottet *et al.*, 2014). The volatile fatty acids (VFA) content (acetate C2, propionate C3, iso-butyrate iC4, butyrate C4, iso-valerate iC5 and valerate C5) was measured in the soluble phase using a gas chromatography (GC-800 Fisons Instrument) equipped with a flame ionisation detector (Mottet *et al.*, 2014). Lactate was analyzed by HPLC (Quéméneur *et al.*, 2011).

6.2.5 Calculations

The sulphate concentrations was expressed as S-SO$_4^{2-}$ using Eq. 6-1 and the S^{2-} concentrations, resulted from the methylene blue analysis, was expressed as S-S^{2-}. All C2, C3, iC4, C4, iC5 and C5 concentrations were expressed, respectively, as COD (COD$_{C2-C5}$) and the addition of them is the total VFA expressed as COD concentration (COD_{TVFA}). Lactate was also represented in terms of COD concentration ($COD_{Lactate}$). At the beginning of the experiments (t_0), $COD_{Lactate}$ was the total COD (TCOD) and the only source of COD concentration in the bioreactors, $COD_{TVFA} = 0$. The TCOD is the addition of the COD_{TVFA} concentration (or COD_{C2-C5}) and $COD_{Lactate}$, if present at any other time of the BSR process (t_{0+1}). The fractions of electron donor ($COD_{Lactate}$), acceptor ($S-SO_4^{2-}$), NH_4^+ and products ($S-S^{2-}$ and COD_{C2-C5}) were calculated using the Eq. 6-2, wherein A is $S-SO_4^{2-}$, $S-S^{2-}$, $COD_{Lactate}$, TCOD, COD_{C2-C5}, COD_{TVFA} and NH_4^+ concentration at t_{0+1} and B is $S-SO_4^{2-}$, TCOD and NH_4^+ concentration at t_0. The volumetric and specific rates were calculated according to Eq. 6-3-Eq. 6-4 at the time of evaluation (t_e) corresponding to the steepest slope and using the initial VSS concentration (VSS_{t0}) in the batch bioreactors. For V_r of sulphate, the numerator of Eq. 6-3 should be divided by the factor of sulphur in sulphate (0.3333). The yield of sulphate removed on the TCOD was calculated using Eq. 6-5, in this equation the consumed fraction of sulphate ($1-f_{S-SO42-}$) at t_e was multiplied by the initial sulphate concentration for the respective batch incubation and further divided by the concentration of TCOD at the same time t_e.

$S-SO_4^{2-}$	$S - SO_4^{2-} = [SO_4^{2-}] \times 0.3333$	Eq. 6-1
Fraction of A (f_A)	$f_A = \dfrac{[A]_{t_{0+1}}}{[B]_{t_0}}$	Eq. 6-2
Volumetric rate (V_r)	$V_r = \dfrac{[1 - (f_A)_{t_e}] \times (B)}{t_e \times \left(\dfrac{1\,d}{24\,h}\right)}$	Eq. 6-3
Specific rate (S_r)	$S_r = \dfrac{V_r}{VSS_{t_0}}$	Eq. 6-4
Yield of sulphate removed on TCOD used	$Y_{SO_4^{2-}/TCOD} = \dfrac{\left(1 - f_{S-SO_4^{2-}}\right)_{t_e} \times SO_4^{2-}}{\left(1 - f_{TCOD}\right)_{t_e} \times TCOD_{t_0}}$	Eq. 6-5

6.3   Results

6.3.1   Anaerobic sulphate reduction at different COD:SO$_4^{2-}$ ratios

In the experiments using electron donor feast conditions and 417 mg $SO_4^{2-}.L^{-1}$ as initial sulphate concentration (COD:SO$_4^{2-}$ ratio of 2.4), the removal of $S-SO_4^{2-}$ fraction showed different slopes (Figure 6-1). Besides, in experiments (A and A*) using electron donor feast conditions (COD:SO$_4^{2-}$ ratio of 2.4), the lactate was removed simultaneously to the amount of sulphate (Figure 6-1A and A*, Figure 6-2A and A*) $r^2 \geq 0.99$. At NH_4^+ famine conditions, a V_r equal to 2,116 mg $SO_4^{2-}.L^{-1}d^{-1}$ was observed and at NH_4^+ feast conditions 1,834 mg $SO_4^{2-}.L^{-1}d^{-1}$ in the course of the BSR, respectively, in A and A* batch bioreactors. The removal fractions of sulphate in the experiments A and A* were, respectively, 0.96 and 0.98 within 8 h of performance (Table 6-2). The V_r of $COD_{Lactate}$ consumption also showed differences in the absence (5,526 mg $COD_{Lactate}.L^{-1}d^{-1}$) and presence (4,140 mg $COD_{Lactate}.L^{-1}d^{-1}$) of initial NH_4^+ concentration, respectively in A and A* batch bioreactors. Furthermore, the V_r of TCOD consumption were 3,448 mg $TCOD.L^{-1}d^{-1}$ and 2,240 mg $TCOD.L^{-1}d^{-1}$ for A and A* respective to the batch bioreactors. $COD_{Lactate}$ was consumed to produce COD_{VFA} like COD_{C2} and COD_{C3} during the BSR (Figure 6-2 and Figure 6-3).

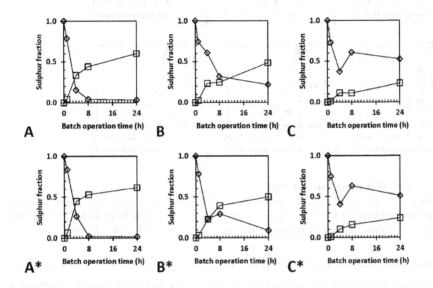

Figure 6-1. Profiles of sulphur species in batch incubations
Fraction of $S-SO_4^{2-}$ ($\Diamond$) and $S-S^{2-}$ ($\square$) in batch incubations. The electron donor feast and famine conditions at different $COD:SO_4^{2-}$ ratios: 2.4 (A and A*), 1.5 (B and B*) and 0.67 (C and C*). The NH_4^+ feast and famine conditions: 0 mg.L^{-1} (A, B and C) and 300 mg.L^{-1} (A*, B* and C*)

The BSR using feast conditions equivalent to a $COD:SO_4^{2-}$ ratio of 1.5 showed larger V_r of removal (4,065 mg $SO_4^{2-}.L^{-1}d^{-1}$, 6,697 mg $COD_{Lactate}.L^{-1}d^{-1}$ and 5,319 mg $TCOD.L^{-1}d^{-1}$) during NH_4^+ famine conditions (B) in comparison to (3,488 mg $SO_4^{2-}.L^{-1}d^{-1}$, 4,920 mg $COD_{Lactate}.L^{-1}d^{-1}$ and 2,921 mg $TCOD.L^{-1}d^{-1}$) NH_4^+ feast conditions (B*) in batch bioreactors. It is unlikely that the larger removal fraction of sulphate was at NH_4^+ feast conditions (0.91 in B*) in comparison to the NH_4^+ famine conditions (0.78 in B) after 24 h of performance (Table 6-2).

The BSR using electron donor famine conditions ($COD:SO_4^{2-}$ ratio of 0.67) showed larger V_r of removal (9,673 mg $SO_4^{2-}.L^{-1}d^{-1}$, 7,033 mg $COD_{Lactate}.L^{-1}d^{-1}$ and 5,715 mg $TCOD.L^{-1}d^{-1}$) during NH_4^+ famine conditions (C) in comparison to (9,019 mg $SO_4^{2-}.L^{-1}d^{-1}$, 5,703 mg $COD_{Lactate}.L^{-1}d^{-1}$ and 3,553 mg $TCOD.L^{-1}d^{-1}$) NH_4^+ feast conditions (C*) in batch bioreactors.

The sulphate removal fractions were very similar (0.48 in C and 0.49 in C*) at NH_4^+ famine or feast conditions, respectively, after 24 h of performance (Table 6-2).

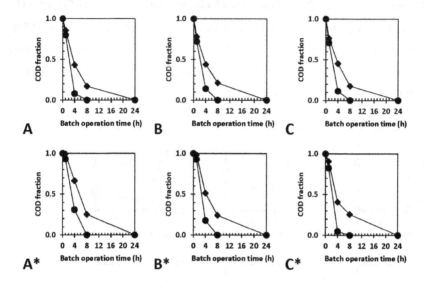

Figure 6-2. Profiles of electron donor (COD)
Profiles of $COD_{Lactate}$ (●) and TCOD (◆) fraction in batch incubations. The electron donor feast and famine conditions at different $COD:SO_4^{2-}$ ratios: 2.4 (A and A*), 1.5 (B and B*) and 0.67 (C and C*). The NH_4^+ feast and famine conditions: 0 mg.L^{-1} (A, B and C) and 300 mg.L^{-1} (A*, B* and C*)

In all the bioreactor incubations lactate was consumed within 8 h and the TCOD was totally consumed within 24 h. In all batch bioreactors, acetate was the major VFA observed after 4 h of incubation, but was also completely consumed within 24 h. A propionate fraction < 0.1 was produced after 4 h and consumed within 8 h in all batch bioreactor incubations, except for experiment A* where propionate was consumed only after 8 h. During the batch incubations at NH_4^+ famine conditions (A, B and C), the NH_4^+ results were always lower than the limit of detection. On the other hand, in the batch incubations at NH_4^+ feast conditions (A*, B* and C*), the NH_4^+ consumption profiles (Figure 6-4) showed similar trends as those of the S-SO_4^{2-} removal fraction (Figure 6-1).

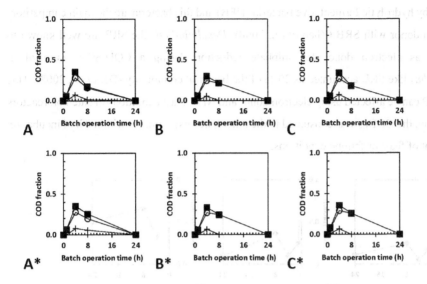

Figure 6-3. Profiles of volatile fatty acids in batch incubations
Profiles of COD_{VFA} (•), COD_{C2} (○) and COD_{C3} (+) fraction in batch incubations. The electron donor feast and famine conditions at different $COD:SO_4^{2-}$ ratios: 2.4 (A and A*), 1.5 (B and B*) and 0.67 (C and C*). The NH_4^+ feast and famine conditions: 0 mg.L^{-1} (A, B and C) and 300 mg.L^{-1} (A*, B* and C*)

6.4   Discussion

6.4.1   Electron donor utilization by sulphate reducing sludge

BSR in bioreactors which operate at low biomass concentration, as biofilm and/or at high rate sulphate feeding conditions, might hamper in a short term operation under nitrogen source famine condition. On the other hand, the reactors operating at high biomass concentration, those as batch, UASB and/or sequencing batch reactors might be more robust in long time operation under feast and famine conditions of a nitrogen source.

This study showed that the sulphate removal improves (> 0.95 sulphate removal fraction, Figure 6-5A) using electron donor feast conditions, e.g. $COD:SO_4^{2-}$ ratio of 2.4, and it is negatively affected using electron donor famine conditions, e.g. $COD:SO_4^{2-}$ ratio of 1.5-0.67. Using lactate in a sulphidogenic sequencing batch bioreactor, the complete sulphate removal efficiency (> 98%) is possible at $COD:SO_4^{2-}$ ratio of 2.4 (Torner-Morales and Buitrón, 2009).

Torner-Morales and Buitrón (Torner-Morales and Buitrón, 2009) showed that sulphate reduction, at $COD:SO_4^{2-}$ ratio of 2.4, can be modelled using only the lactate information. However, propionate formation and consumption was observed in all batch experiments (Figure 6-3). The formation of COD_{C3} (Eq. 6-6) is normally taking place due to the fermentation of

$COD_{Lactate}$ by hydrolytic fermentative bacteria (HFB) and this bacteria are the main competitors for electron donor with SRB (Zhao *et al.*, 2008). *Desulfobulbus* like SRB are well known to utilize C3 as electron donor for sulphate reduction. Using a $COD:SO_4^{2-}$ ratio of 2, *Desulfobulbus* like SRB accounted as 20 % of the bacterial community (Dar *et al.*, 2008). This type of SRB can be affected using electron donor and NH_4^+ feast conditions in batch bioreactors A*. Unlikely, the COD_{C3} was consumed in the other batch experiments within 8 h of incubation independent of feast or famine conditions.

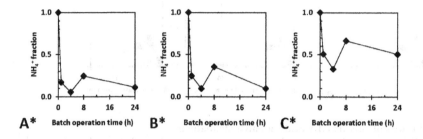

Figure 6-4. Profiles of NH_4^+ (◆) fraction in batch incubations
The electron donor feast and famine conditions at different $COD:SO_4^{2-}$ ratios: 2.4 (A*), 1.5 (B*) and 0.67 (C*). The NH_4^+ feast and famine conditions: 0 mg.L^{-1} (A, B and C) and 300 mg.L^{-1} (A*, B* and C*)

Two main processes could be inferred from the slopes of the $S-SO_4^{2-}$ profiles, the simultaneous evolution of COD_{VFA} production and $COD_{Lactate}$ consumption. i) The sulphate removal (Figure 6-1) by means of $COD_{Lactate}$ utilization was the first process (Eq. 6-7 to Eq. **6-8**, Table 6-3). Meanwhile, a small fraction of C3 (< 0.1) was produced by the HFB (Eq. 6-6) during the competition with SRB for $COD_{Lactate}$ fermentation, followed by COD_{C3} utilization by the SRB (Eq. 6-9 and Eq. 6-10). Secondly, ii) COD_{C3} was produced from $COD_{Lactate}$ by the HFB at high rates (Eq. 6-6), while the COD_{C3} was consumed simultaneously by SRB (Eq. 6-9 and Eq. 6-10) at an approximately similar rate as it was produced by the HFB. Both processes result in C2 formation. Acetotrophic sulphate reduction occurred at electron donor famine conditions and high concentrations of sulphate (1491 mg SO_4^{2-}. L^{-1}), experiments C and C*. According to the Gibbs free energy of formation, the biochemical pathway ii (Eq. 6-6, Eq. **6-9** and Eq. **6-10**, Table 6-3) is the most preferred. However, the δ-proteobacteria was the most abundant group rather than firmicutes during sulphate reduction using lactate as electron donor (Zhao *et al.*, 2010), this result suggests that the biochemical pathway i (Eq. 6-7 to Eq. **6-10**) was the dominant in the literature.

Table 6-2. Parameters used for kinetic evaluation during feast or famine initial conditions in batch bioreactors

	COD:SO$_4^{2-}$ ratio	Batch	Fraction removed of SO$_4^{2-}$(time of evaluation)	Fraction removed of TCOD (time of evaluation)	Y$_{SO42-/COD}$ (mg SO$_4^{2-}$. mg TCOD^{-1})	Pearson correlation value of lactate utilization on SO$_4^{2-}$ (r^2)
NH$_4^+$ Famine (0 mg. L^{-1})						
Electron donor feast	2.4	A	0.96 (8)	0.83 (8)	0.49	1.0 (r^2=1.0)
to	1.5	B	0.78 (24)	1.0 (24 h)	0.52	
Electron donor famine	0.67	C	0.48 (24)	1.0 (24 h)	0.71	
NH$_4^+$ Feast (300 mg. L^{-1})						
Electron donor feast	2.4	A*	0.98 (8 h)	0.75 (8 h)	0.55	1.0 (r^2=0.99)
to	1.5	B*	0.91(24)	1.0 (24 h)	0.61	
Electron donor famine	0.67	C*	0.49 (24)	1.0 (24 h)	0.73	

Note: *denotes the tests under NH$_4^+$ feast coinditions

In experiments C and C*, acetotrophic sulphate reduction can be explained by the final sulphate concentrations removed rather than the fraction removed, larger amount of sulphate was removed under electron donor famine than in feast conditions: e.g. 715 and 730 mg SO$_4^{2-}$.L^{-1} were removed, respectively, in experiment C and C*at electron donor famine conditions, in comparison to 400 and 408 mg SO$_4^{2-}$. L^{-1} removed during the experiments A and A*, respectively, at electron donor feast conditions. Notwithstanding, acetotrophic sulphate reduction (Eq. 6-11) is not supported by the -48 kJ.reaction^{-1} Gibbs free energy of formation compared to the -31.1 kJ.reaction^{-1} Gibbs free energy of formation for acetotrophic methanogenesis (Eq. 6-14). Besides, SRB have a lower affinity (> K_S) for C2 compared to acetotrophic methanogenic archaea (< K_S) (Stams et al., 2005). In contrast, at low COD:SO$_4^{2-}$ ratios, SRB can outcompete methanogenic archae for acetotrophic sulphate reduction (Chou et al., 2008).

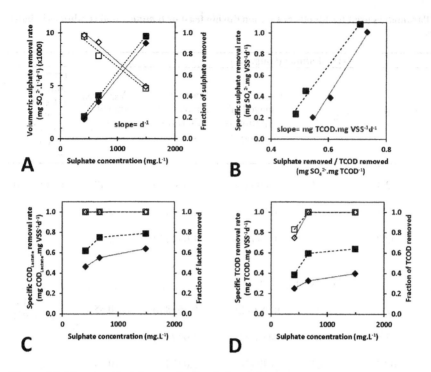

Figure 6-5. Kinetics of sulphate reduction in batch bioreactors
A) The volumetric sulphate removal rate and the best fraction of sulphate removed against the initial sulphate concentration. B) The specific sulphate removal rate compared to the experimental yield of sulphate removed on TCOD. C) The specific COD$_{Lactate}$ removal rate and fraction of lactate removed after 8 h of operation as function of initial sulphate concentration. D) The specific TCOD removal rate and fraction of TCOD removed at the end of the sulphate reduction process as function of initial sulphate concentration. In the primary axis NH$_4^+$ feast ($\blacklozenge$) and famine ($\blacksquare$) conditions. In the secondary axis NH$_4^+$ feast ($\lozenge$) and famine ($\square$) conditions

Table 6-3. Stoichiometric reactions in batch bioreactors

Reaction	$\Delta G^{0\prime}$ (kJ/reaction)	Eq. No.
3 Lactate$^-$ $\rightarrow$ 2 Propionate$^-$ + Acetate$^-$ + CO$_2$	-54.9	Eq. 6-6
2 Lactate$^-$ + SO$_4^{2-}$ $\rightarrow$ 2 Acetate$^-$ + HS$^-$ + 2HCO$_3^-$ + H$^+$	-160.1	Eq. 6-7
2Lactate$^-$ + 3SO$_4^{2-}$ $\rightarrow$ 6 HCO$_3^-$ + HS$^-$ + H$^+$	-255.3	Eq. 6-8
Propionate$^-$ + 0.75 SO$_4^{2-}$ $\rightarrow$ Acetate$^-$ + HCO$_3^-$ + 0.75 HS$^-$ + 0.25 H$^+$	-37.7	Eq. 6-9
Propionate$^-$ + 1.75 SO$_4^{2-}$ $\rightarrow$ 3 HCO$_3^-$ + 1.75 HS$^-$ + 0.25 H$^+$	-85.4	Eq. 6-10
Acetate$^-$ + SO$_4^{2-}$ $\rightarrow$ 2 HCO$_3^-$ + HS$^-$	-48	Eq. 6-11
4 H$_2$ + SO$_4^{2-}$ + H$^+$ $\rightarrow$ HS$^-$ + 4 H$_2$O	-151.9	Eq. 6-12
4 H$_2$ + HCO$_3^-$ + H$^+$ $\rightarrow$ CH$_4$ + 3 H$_2$O	-135.6	Eq. 6-13
Acetate$^-$ + H$_2$O $\rightarrow$ CH$_4$ + HCO$_3^-$	-31.1	Eq. 6-14

6.4.2 Kinetics of sulphate reduction under feast and famine conditions in batch

Regardless of the electron donor and NH_4^+ feast or famine condition, the sulphate removal rate improved at increasing the initial concentrations of sulphate (Figure 6-5A). According to the Eq. 6-15, the sulphate removal process followed a kinetic of first order (k_1).

First order kinetic equation
$$-\frac{dS}{dt} = k_1 S$$
Eq. 6-15

The Eq. 6-15 is recommended for the kinetic evaluation of substrate consumption in non-supporting growth conditions (Schmidt *et al.*, 1985). In this research, the cell growth was considered negligible due to the initial substrate to biomass ratio. The k_1 of sulphate removal was 4 % lower using NH_4^+ feast conditions, k_1=6.69 d^{-1} (r^2=1), compared to NH_4^+ famine conditions, k_1=6.98 d^{-1} (r^2=0.9992), Figure 6-5A. The high initial sulphate concentration (1491 mg $SO_4^{2-}.L^{-1}$) was not in the range to inhibit the sulphate removal process, according to the literature (Bernardez *et al.*, 2013). The fractions of sulphate removed was severely affected by a deficiency of electron donor (famine conditions) and dropped from > 0.95 to < 0.5 (Table 6-2).

During the BSR, the specific electron donor consumption rate (mg TCOD.mg VSS^{-1}d^{-1}) was 16.6 % larger during NH_4^+ feast conditions, 4.39 mg TCOD.mg VSS^{-1}d^{-1} (r^2 = 0.9895), compared to NH_4^+ famine conditions, 3.66 mg TCOD.mg VSS^{-1}d^{-1} (r^2 = 0.99), Figure 6-5B. NH_4^+ feast conditions could drive to anabolic pathways, e.g formation of new biomass, rather than the bioprocess of nitrification and denitrification. In this research, neither nitrate nor nitrite were detected in the soluble phase. Besides, the NH_4^+ profiles followed similar trends of consumptions as those of sulphate. During BSR and NH_4^+ famine conditions, *Desulfovibrio desulfuricans* decreased the electron donor uptake, the cell size was negatively affected and the cell carbon content decreased (Okabe *et al.*, 1992). Additionally, the electron donor (COD$_{Lactate}$ and/or TCOD) removal rates were faster during NH_4^+ famine compared to NH_4^+ feast conditions and are also affected by the initial also sulphate concentration in the batch incubations (Figure 6-5C-D).

Anabolic pathways might result in lower specific rates when compared to degradation rates, therefore it might represent a type of limiting process, which seems to be the case for NH_4^+ feast in this research. For instance, in BSR experiments with *Archaeoglobus fulgidus* Strain Z and sulphate feast conditions, biomass formation on sulphate consumption was higher compared to sulphate starving conditions (Habicht *et al.*, 2005). Also carbon uptake for biomass formation was greater at sulphate feast conditions. The nitrification and denitrification process is developed at higher specific rates in comparison to sulphate reduction process. In principle, nitrification and denitrification are biochemical pathways that take place in the periplasmic space by a chain of enzymatic systems. In contrast, the transport of sulphate to the cytoplasm and the amount of ATP which should be spent to reduce it. Most likely, cells consume electron donors and will choose for the thermodynamically optimal pathway, *e.g.* a shift of the biochemical pathways due to sulphate limiting and non-limiting conditions (Habicht *et al.*, 2005).

6.5   Conclusions

Biomass performs sulphate reduction optimally under feast conditions of electron donor, the sulphate removal efficiency was > 95 %. The first order kinetic constant (k_I) of sulphate removal was 4 % lower using NH_4^+ feast conditions and compared to famine conditions. The specific electron donor consumption rate, 4.39 mg TCOD.mg $VSS^{-1}d^{-1}$ (r^2=0.9895), improved 16.6 % using NH_4^+ feast conditions, compared to NH_4^+ famine conditions, 3.66 mg TCOD. mg VSS^{-1} d^{-1} (r^2=0.99). NH_4^+ feast or famine conditions influenced the metabolic pathways used by the sulphate reducing sludge. Besides, the electron donor flow to sulphate reduction improved at increasing electron donor famine conditions.

6.6   References

APHA, 1999. Standard Methods for the Examination of Water and Wastewater, 20th ed. Washington, USA.

Bernardez, L. A., de Andrade Lima, L.R.P., de Jesus, E.B., Ramos, C.L.S., Almeida, P.F., 2013. A kinetic study on bacterial sulfate reduction. Bioprocess Biosyst. Eng. 36, 1861–1869. doi:10.1007/s00449-013-0960-0

Chou, H.-H., Huang, J.-S., Chen, W.-G., Ohara, R., 2008. Competitive reaction kinetics of sulfate-reducing bacteria and methanogenic bacteria in anaerobic filters. Bioresour. Technol. 99, 8061–8067. doi:10.1016/j.biortech.2008.03.044

Dar, S. A., Kleerebezem, R., Stams, A.J.M., Kuenen, J.G., Muyzer, G., 2008. Competition and coexistence of sulfate-reducing bacteria, acetogens and methanogens in a lab-scale anaerobic bioreactor as affected by changing substrate to sulfate ratio. Appl. Microbiol. Biotechnol. 78, 1045–1055. doi:10.1007/s00253-008-1391-8

Habicht, K.S., Salling, L., Thamdrup, B., Canfield, D.E., 2005. Effect of low sulfate concentrations on lactate oxidation and isotope fractionation during sulfate reduction by *Archaeoglobus fulgidus* strain Z. Appl. Environ. Microbiol. 71, 3770–3777. doi:10.1128/AEM.71.7.3770-3777.2005

Kaksonen, A.H., Puhakka, J.A., 2007. Sulfate reduction based bioprocesses for the treatment of acid mine drainage and the recovery of metals. Eng. Life Sci. 7, 541–564. doi:10.1002/elsc.200720216

Lens, P., Vallerol, M., Esposito, G., Zandvoort, M., 2002. Perspectives of sulfate reducing bioreactors in environmental biotechnology. Rev. Environ. Sci. Bio/Technology 1, 311–325. doi:10.1023/A:1023207921156

Lewis, A.E., 2010. Review of metal sulphide precipitation. Hydrometallurgy 104, 222–234. doi:10.1016/j.hydromet.2010.06.010

Liamleam, W., Annachhatre, A.P., 2007. Electron donors for biological sulfate reduction. Biotechnol. Adv. 25, 452–463. doi:10.1016/j.biotechadv.2007.05.002

Mizuno, O., Li, Y.Y., Noike, T., 1998. The behavior of sulfate-reducing bacteria in acidogenic phase of anaerobic digestion. Water Res. 32, 1626–1634. doi:10.1016/S0043-1354(97)00372-2

Mottet, A., Habouzit, F., Steyer, J.P., 2014. Anaerobic digestion of marine microalgae in different salinity levels. Bioresour. Technol. 158, 300–306. doi:10.1016/j.biortech.2014.02.055

Okabe, S., Nielsen, P.H., Charcklis, W.G., 1992. Factors affecting microbial sulfate reduction by *Desulfovibrio desulfuricans* in continuous culture: limiting nutrients and sulfide concentration. Biotechnol. Bioeng. 40, 725–734. doi:10.1002/bit.260400612

Papirio, S., Villa-Gomez, D.K., Esposito, G., Pirozzi, F., Lens, P.N.L., 2013. Acid mine drainage treatment in fluidized-bed bioreactors by sulfate-reducing bacteria: A Critical Review. Crit. Rev. Environ. Sci. Technol. 43, 2545–2580. doi:10.1080/10643389.2012.694328

Plugge, C.M., Zhang, W., Scholten, J.C.M., Stams, A.J.M., 2011. Metabolic flexibility of sulfate-reducing bacteria. Front. Microbiol. 2, 1–8. doi:10.3389/fmicb.2011.00081

Quéméneur, M., Hamelin, J., Latrille, E., Steyer, J.-P., Trably, E., 2011. Functional versus phylogenetic fingerprint analyses for monitoring hydrogen-producing bacterial populations in dark fermentation cultures. Int. J. Hydrogen Energy 36, 3870–3879. doi:10.1016/j.ijhydene.2010.12.100

Schmidt, S.K., Simkins, S., Alexander, M., 1985. Models for the kinetics of biodegradation of organic compounds not supporting growth. Appl. Environ. Microbiol. 50, 323–31.

Stams, A.J.M., Plugge, C.M., de Bok, F.A.M., van Houten, B.H.G.W., Lens, P., Dijkman, H., Weijma, J., 2005. Metabolic interactions in methanogenic and sulfate-reducing bioreactors. Water Sci. Technol. 52, 13–20.

Thauer, R.K., Jungermann, K., Decker, K., 1977. Energy conservation in chemotrophic anaerobic bacteria. Bacteriol. Rev. 41, 100–180. doi:10.1073/pnas.0803850105

Torner-Morales, F.J., Buitrón, G., 2009. Kinetic characterization and modeling simplification of an anaerobic sulfate reducing batch process. J. Chem. Technol. Biotechnol. 85, 453–459. doi:10.1002/jctb.2310

Villa-Gomez, D.K., Papirio, S., van Hullebusch, E.D., Farges, F., Nikitenko, S., Kramer, H., Lens, P.N.L., 2012. Influence of sulfide concentration and macronutrients on the characteristics of metal precipitates relevant to metal recovery in bioreactors. Bioresour. Technol. 110, 26–34. doi:10.1016/j.biortech.2012.01.041

Zhao, Y., Ren, N., Wang, A., 2008. Contributions of fermentative acidogenic bacteria and sulfate-reducing bacteria to lactate degradation and sulfate reduction. Chemosphere 72, 233–242. doi:10.1016/j.chemosphere.2008.01.046

Zhao, Y.-G., Wang, A.-J., Ren, N.-Q., 2009. Effect of sulfate absence and nitrate addition on bacterial community in a sulfidogenic bioreactor. J. Hazard. Mater. 172, 1491–1497. doi:10.1016/j.jhazmat.2009.08.018

Zhao, Y.-G., Wang, A.-J., Ren, N.-Q., 2010. Effect of carbon sources on sulfidogenic bacterial communities during the starting-up of acidogenic sulfate-reducing bioreactors. Bioresour. Technol. 101, 2952–2959. doi:10.1016/j.biortech.2009.11.098

Chapter 7

The effect of feast and famine conditions on biological sulphate reduction in anaerobic sequencing batch reactors

Abstract

Parameters affecting the sulphate removal during the biological treatment of industrial wastewater are not well known and might hamper the biological system. This research shows the effect of transient feeding conditions on the sulphate removal using sequencing batch reactors (SBR). Six SBR were used, R1L (1 g $COD_{Lactate}.L^{-1}$, 0.4 g $SO_4^{2-}.L^{-1}$ and 0.3 g $NH_4Cl.L^{-1}$) and R1H (6 g $COD_{Lactate}.L^{-1}$ and 2.5 g $SO_4^{2-}. L^{-1}$ and 1.8 g $NH_4Cl.L^{-1}$) were used as control and steady feeding. The SBR R2L and R3L operated at feast (6 g $COD_{Lactate}.L^{-1}$ and 2.5 g $SO_4^{2-}.L^{-1}$) and famine conditions (zero concentration). The SBR R2H and R3H operated at higher feast concentrations (36 g $COD_{Lactate}.L^{-1}$ and 15 g $SO_4^{2-}. L^{-1}$) and also famine conditions (zero concentration). The SBR R2L and R2H were feed with 1.8 g $NH_4Cl.L^{-1}$ meanwhile, the SBR R3L and R3H with 0 g $NH_4Cl.L^{-1}$, during the feast conditions. The sulphate removal efficiency (*SRE*) was robust to transient feeding conditions in R2L and R3L, it was 92% at feast and 86-90 % at famine conditions in both SBR while 94 ($\pm$ 8)% at the control reactor R1H. The *SRE* was 79 ($\pm$ 17)% in R1L, dropped to < 20% in R2H and it was maintained at slightly above 40% in R3H.

Keywords: Biological sulphate reduction, industrial wastewater, sulphate reducing bacteria, sequencing batch reactor, transient feeding conditions, feast-famine conditions

7.1 Introduction

Industrial wastewaters containing high sulphate concentrations can dramatically damage the flora and fauna of water reservoir when they are not treated efficiently. Industrial wastewater with high sulphate concentrations is characterized by low pH, high oxidative potential, can contain high concentrations of toxic metals and they lack chemical oxygen demand (COD) (Mapanda *et al.*, 2007). The lack of COD in industrial wastewater rich in sulphate represents an economic disadvantage for the biological treatment. Hence, the use of pure chemicals as electron donors (COD) increases the cost of the biological treatment.

The sulphate removal by sulphate reducing bacteria (SRB) has been studied in different anaerobic reactors as *e.g.* batch reactor (Al-Zuhair *et al.*, 2008), sequencing batch reactor (Torner-Morales and Buitrón, 2010), upflow anaerobic sludge blanket reactor, (UASB) (Bertolino *et al.*, 2012), extended granular sludge bed reactor (EGSB) (Dries *et al.*, 1998) and gas lift reactor (Sipma *et al.*, 2007). Among these studies, factors as COD:sulphate ration and hydrodynamic conditions (effect of the HRT) have been extensively studied for high sulphate removal efficiencies (*SRE*). Transient conditions might hamper the biological sulphate reduction (abnormal bioreactors operation or changes on the wastewater compositions). Nevertheless, the effect of transient conditions on the robustness and resilience of bioreactors performing sulphate removal is not well known. For example, the robustness of a sulphate reducing bioreactor might be defined by the tolerance to transient feeding conditions and the simultaneous capability to perform at steady sulphate removal efficiency, as before the changes in the influent composition. The resilience should be the capability of a sulphate reducing bioreactor to overcome a transient condition and restore the performance as before the changes in the influent composition. The robustness and resilience depends on the syntrophism existing in a biological system (Alon *et al.*, 1999; Barkai and Leibler, 1997). Anaerobic sludge offers the advantage of different bacteria carrying out multiple syntrophic degradation pathways (Whiteley and Lee, 2006). Furthermore, anaerobic sludge is an excellent source of sulphate reducing bacteria (SRB) (Reyes-Alvarado *et al.*, 2017). The SRB are capable to use diverse low molecular weight organic compounds as electron donors, also, they are able to perform heterotrophicaly and autotrophicaly while sulphate removal (Plugge *et al.*, 2011). The biological sulphate reduction in continuous inverse fluidized bed bioreactors (IFBB) is robust and resilient to transient feeding conditions (Reyes-Alvarado *et al.*, 2017). Reyes-Alvarado *et al.* (2017) showed that a sulphidogenic IFBB (experimental IFBB) was robust to transient feeding conditions (feast: 1495 mg $SO_4^{2-}.L^{-1}$, 2133 mg $COD.L^{-1}$ and famine: 0 mg $SO_4^{2-}.L^{-1}$, 0 mg $COD.L^{-1}$) at a sulphate removal efficiency (RE) of 67% and this performance was comparable to that at steady feeding conditions, sulphate RE = 61% (control IFBB) and 71% (experimental IFBB at steady feeding conditions).

The effect of nitrogen source is poorly studied as limiting growth factor for SRB during the sulphate removal. Cell growth needs very important elements as N, P, C and S, this is one of the links between their respective biogeochemical cycles. Bacteria needs amino acids (*e.g.* cytein, methionine, etc.), highly energy loaded molecules (*e.g.* ATP and acetyl-CoA), proteins and DNA in order to accomplish their biological activities. Okabe *et al.* (1992) studied the physiology and performance of *Desulfovibrio desulfuricans* whereas sulphate reduction at increasing nitrogen limiting conditions, the results were: the electron donor uptake decreased, the cell size was negatively affected and the cell carbon composition was diminish.

Therefore, the aim of this research was to study the effect of transient feeding conditions at high sulphate concentrations (2.5 and 15 g $SO_4^{2-}.L^{-1}$) on the biological sulphate removal. The robustness and resilience was studied at no limiting COD:Sulphate ratio (2.4) in sequencing batch bioreactors (SBR). The use of a less dynamic bioreactor as SBR offers higher biomass retention in comparison to a continuous IFBB. Furthermore, the effect of NH_4^+, as nitrogen source, on the sulphate removal was investigated during the transient feeding conditions.

7.2 Material and methods

7.2.1 The sulphate reducing biomass

The sulphate reducing biomass was obtained from two sequencing batch reactors (SBR) performing steady state biological sulphate removal. One reactor was operated at low concentrations, 1 g $COD_{Lactate}.L^{-1}$ and 0.4 g $SO_4^{2-}.L^{-1}$. The second SBR was operated at high concentrations of 6 g $COD_{Lactate}.L^{-1}$ and 2.5 g $SO_4^{2-}.L^{-1}$. Both reactors were operated at a COD:Sulphate ratio of 2.4 and resulted in sulphate RE $\geq$ 90 %.

Three SBR, R1L, R2L and R3L, were inoculated with 2 L containing 7.7 g $VSS.L^{-1}$ of volatile suspended solids (VSS) as sulphate reducing biomass from the SBR operated at low sulphate concentrations. Three other SBR, R1H, R2H and R3H, were inoculated with 2 L containing 15.9 g $VSS.L^{-1}$ of sulphate reducing biomass from the SBR operated at high sulphate concentrations.

7.2.2   Synthetic wastewater

The composition of the synthetic wastewater used in this research was as follows (in mg.L^{-1}): NH$_4$Cl (300), MgCl$_2$•6H$_2$O (120), KH$_2$PO$_4$ (200), KCl (250), CaCl$_2$•2H$_2$O (15), yeast extract (20) and 0.5 mL of a mixture of micronutrients. The micronutrients solution content (in mg.L^{-1}): FeCl$_2$•4H$_2$O (1500), MnCl$_2$•4H$_2$O (100), EDTA (500), H$_3$BO$_3$ (62), ZnCl$_2$ (70), NaMoO$_4$•2H$_2$O (36), AlCl$_3$•6H$_2$O (40), NiCl$_3$•6H$_2$O (24), CoCl$_2$•6H$_2$O (70), CuCl$_2$•2H$_2$O (20) and HCl 36 % (1 mL) (Villa-Gomez et al., 2012). Sodium lactate and sodium sulphate were used as electron donor and acceptor, respectively, and maintained a COD:Sulphate ratio of 2.4 at any circumstance. The source of nitrogen (300 mg NH$_4$Cl.L^{-1}) was excluded from the preparation of the synthetic wastewater when it was needed by the experiment. The synthetic wastewater influent pH was adjusted to 6.0 with NaOH (1 M) or HCl (1 M). All reagents used in this study were of analytical grade.

7.2.3   Reactor set up

Six SBR of 2 L active volume were configured with a stirring system and consisted of an axis with two propels, the speed was fixed at 120 rpm. The SBR were fed and discharged with two peristaltic pumps (Masterflex L/S). The temperature was controlled at 30 °C, with a water bath (Cole Parmer, Polystat 12112-00). Figure 7-1 describes the configuration of the six SBR. The SBR operated 3 cycles of 8 h per day. The schedule of each cycle was as follows: 2 minutes to discharge, two minutes of feeding, 3 hours of agitation stating from time zero and 5 hours of settling from 3 to 8 h of the cycle. 0.33 L of supernatant was removed and the same volume was fed, resulting in a liquid flow in (Q$_{in}$) of 1 L. d^{-1} or an HRT=2 d.

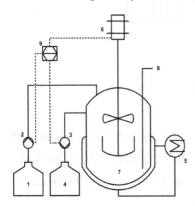

Figure 7-1. Schematic of a SBR
Components: 1) Influent tank, 2) Peristaltic pump, 3) Peristaltic pump, 4) Effluent tank, 5) Heating system (water bath), 6) Stirring system, 7) SBR, 8) Sampling port and 9) Timer

7.2.4 Transient, feast-famine, feeding conditions in SBR operation

Two SBR were used as control experiments, R1L and R1H. The SBR R1L used low concentrations: 1 g $COD_{Lactate}.L^{-1}$ and 0.4 g $SO_4^{2-}.L^{-1}$ (140 mg $S-SO_4^{2-}.L^{-1}$). The SBR R1H performed at high concentrations: 6 g $COD_{Lactate}.L^{-1}$ and 2.5 g $SO_4^{2-}.L^{-1}$ (840 mg $S-SO_4^{2-}.L^{-1}$). Neither the COD or sulphate concentrations nor the HRT were modified during the 22 d of SBR operation.

The SBR R2L and R3L were operated at transient feeding conditions. Both reactors followed the schedule described in Figure 7-2. The first four days, both SBR were operated at steady feeding conditions at 1 g $COD_{Lactate}.L^{-1}$ and 0.4 g $SO_4^{2-}.L^{-1}$ (140 mg $S-SO_4^{2-}.L^{-1}$), followed by a famine day. In a famine day, the influent pump stopped and zero electron donor and acceptor were fed. The famine operation was followed by a feast day. During a feast day, 6 g $COD_{Lactate}.L^{-1}$ and 2.5 g $SO_4^{2-}.L^{-1}$ (0.84 g $S-SO_4^{2-}.L^{-1}$) were fed. The SBR R2H and R3H were also operated at transient feeding conditions (Figure 7-2) but at higher concentrations. The first four days, both SBR were operated at steady feeding conditions at 6 g $COD_{Lactate}.L^{-1}$ and 2.5 g $SO_4^{2-}.L^{-1}$ (0.84 g $S-SO_4^{2-}.L^{-1}$), and further followed by a famine day. The famine operation was followed by a feast day but at 36 g $COD_{Lactate}.L^{-1}$ and 15 g $SO_4^{2-}.L^{-1}$ (5 g $S-SO_4^{2-}.L^{-1}$).

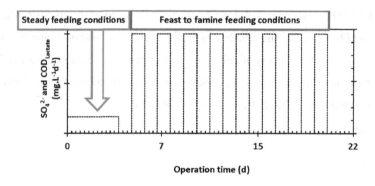

Figure 7-2. Schedule of transient feeding conditions in SBR

The SBR R1L operated at constant NH_4^+ concentrations (300 mg $NH_4Cl.L^{-1}$). For the SBR: R2L, R1H and R2H, the NH_4^+ source (NH_4Cl) was increased 6 times (1800 mg $NH_4Cl.L^{-1}$) during steady and transient feeding operation. The NH_4^+ source (300 mg $NH_4Cl.L^{-1}$) was only excluded from the influent feeding for SBR R3L and R3H.

7.2.5 Evaluation of the performance of SBR

The reactor activity was evaluated in terms of the loading rates (*LR*, Eq. 7-1) compared to the removal rates (*RR*, Eq. 7-2), fraction (*f*, Eq. 7-3) of a component in the effluent, removal efficiencies (*RE*, Eq. 7-4). The robustness of the process was evaluated by comparison of control SBR (R1L and R1H) and experimental SBR (R2L, R3L, R2H and R3H).

$$LR = \frac{Q(A)}{V} \qquad\qquad \text{Eq. 7-1}$$

$$RR = \frac{Q(A-B)}{V} \qquad\qquad \text{Eq. 7-2}$$

$$f = \frac{(A-B)}{A} \qquad\qquad \text{Eq. 7-3}$$

$$RE = \left(\frac{(A-B)}{A}\right) * 100 \qquad\qquad \text{Eq. 7-4}$$

The flow rate (Q) of 1 L.d^{-1} was equivalent to 3 cycles.d^{-1}. The operational volume (V) of the SBR was expressed in L. The n-compound influent concentration (A) to the SBR, the n-compound effluent concentration (B) was expressed in mg.L^{-1}, *e.g.* compounds like the initial SO_4^{2-}, S- SO_4^{2-}, lactate or total COD (TCOD) concentrations. The TCOD was calculated as the addition of volatile fatty acids (VFA, mg COD.L^{-1}) and lactate (mg COD.L^{-1}).

7.2.6 Chemical analysis

The VSS, total suspended solids (TSS) and sulphide (total dissolved sulfide by the methylene blue reaction) were measured according to the procedure outlined in Standard Methods (APHA, 1999). The pH was measured off-line with a sulphide resistant electrode (Prosense, Oosterhout, The Netherlands). Sulphate and ammonia were analyzed by ion chromatography (DIONEX 100) using a conductivity detector, as reported by (Mottet *et al.*, 2014). The VFA content (acetate C2, propionate C3, iso-butyrate iC4, butyrate C4, iso-valerate iC5 and valerate C5) was measured in the soluble phase using a gas chromatography (GC-800 Fisons Instrument) equipped with a flame ionization detector (Mottet *et al.*, 2014). Lactate was analyzed by HPLC as reported in the literature (Quéméneur *et al.*, 2011).

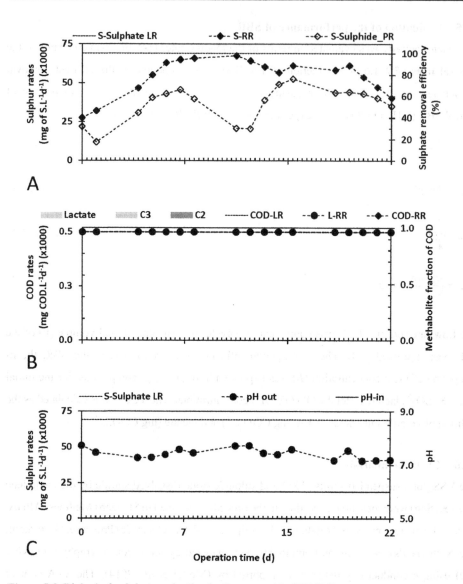

Figure 7-3. Biological sulphate reduction in the anaerobic SBR R1L

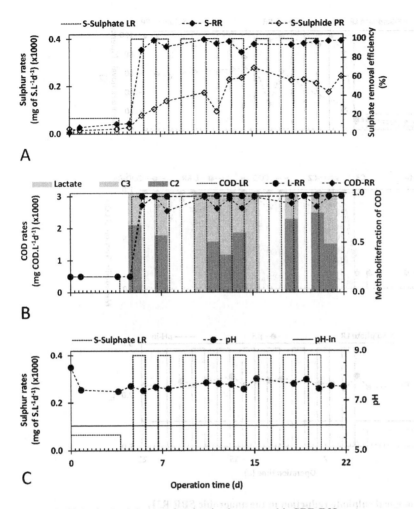

Figure 7-4. Biological sulphate reduction in the anaerobic SBR R2L

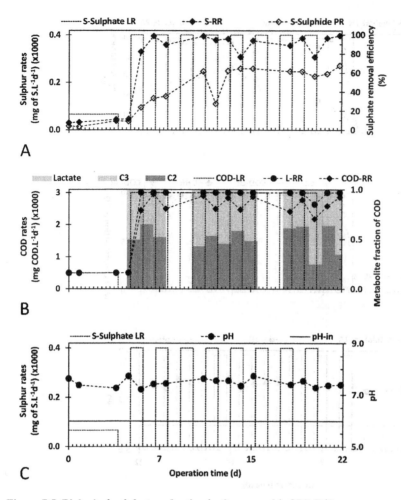

Figure 7-5. Biological sulphate reduction in the anaerobic SBR R3L

7.3 Results

7.3.1 Anaerobic sulphate reduction in SBR at steady feeding conditions

Figure 7-3A-C describes the performance of SBR R1L, the SBR flow rates in and out are shown and expressed on a daily basis. Table 7-1 reports the average, maximum and minimum values observed during the operation of the SBR under all conditions tested. When the $S-SO_4^{2-}-RR$ reached a value equal to the $S-SO_4^{2-}-LR$, the sulphate RE was assumed 100%. The reactor showed 79 ($\pm$ 17)% sulphate RE and 100% of COD and lactate RE. In this SBR, no VFA was detected in the R1L effluent and the pH was always above the neutral, $e.g.$ 7.4 ($\pm$ 0.2) (Figure 7-3C).

Figure 7-6A-C shows the performance of SBR R1H that achieved 94 ($\pm$ 8)% of sulphate RE, lactate RE 99 ($\pm$ 3)% and COD RE of 93 ($\pm$ 7)%. Acetate and propionate were the most abundant fraction of VFA in the COD effluent composition. The pH was always above the neutral, $e.g.$ 7.7 ($\pm$ 0.2) Figure 7-6C.

7.3.2 Anaerobic sulphate reduction in SBR at transient feeding conditions

The SBR R2L performed sulphate removal at transient feeding conditions, feast 92 ($\pm$ 13)% and famine 90 ($\pm$ 11)% sulphate RE. Lactate was always consumed at 100 % RE and the COD RE differed between the feast 99 ($\pm$ 1)% and famine 92 ($\pm$ 6)% conditions. The pH was 7.7 ($\pm$ 0.1) during feast and 7.5 ($\pm$ 0.1) during the famine conditions (Figure 7-4A-C).

The SBR R2H was fed with high concentrations (36 g $COD_{Lactate}.L^{-1}$ and 15 g $SO_4^{2-}. L^{-1}$ or 5 g $S-SO_4^{2-}. L^{-1}$) during the feast periods. Sulphate RE amounted to 55 ($\pm$ 19)% during the feast and 36 ($\pm$ 29)% during the famine conditions. The lactate was not totally consumed, 85 ($\pm$ 8)% RE was observed during the feast and 86 ($\pm$ 14)% during the famine conditions. The COD RE averaged 42 ($\pm$ 26)% during the feast and 35 ($\pm$ 35)% during the famine conditions. In the performance profiles (Figure 7-7A-C), the S-RR (S-SO_4^{2-} RR) and the COD-RR decreased until the end of the experiments. On the other hand, the lactate RR profile recovered after the famine days. The pH averaged 7.2 ($\pm$ 0.3) during the feast and 7.0 ($\pm$ 0.3) during the famine conditions, sometimes the pH decreased to lower than the neutral values.

7.3.3 Anaerobic sulphate reduction in SBR at transient feeding conditions in the absence of NH_4^+

Figure 7-5A-C shows the performance of sulphate removal, lactate and COD consumption at transient feeing conditions when NH_4^+ is excluded from the influent in SBR R3L. The sulphate RE was 92 ($\pm$ 11)% at feast and 86 ($\pm$ 11)% at famine conditions, while the lactate RE was 100 % in both cases. The COD RE was higher (96 $\pm$ 3%) during feast in comparison to famine (85 $\pm$ 8%) conditions. The pH was never below the neutral values, $e.g.$ 7.6 ($\pm$ 0.1) at feast and 7.4 ($\pm$ 0.1) at famine.

Figure 7-8A-C shows the performance of sulphate removal, lactate RR, COD RR and pH during transient feeding conditions without NH_4^+ in the SBR R3H. The sulphate RE was 57 ($\pm$ 21)% at feast and 55 ($\pm$ 28)% at famine conditions. The sulphate RR did not decrease to lower than 30% and the RR recovered from ~ 30 to ~ 43% during the last two famine operation days (Figure 7-8A). Also, the lactate RE was very similar during the feast and famine conditions, 93 ($\pm$ 8)% and 92 ($\pm$ 10)%, respectively. But the COD was not totally consumed, the RE was 37 ($\pm$ 26)%

and 33 (± 34)% for feast and famine conditions, respectively. In this experiment, the pH did not reach values lower than 7.0; the average pH was 7.7 (± 0.3) during the feast and 7.8 (± 0.3) during the famine conditions.

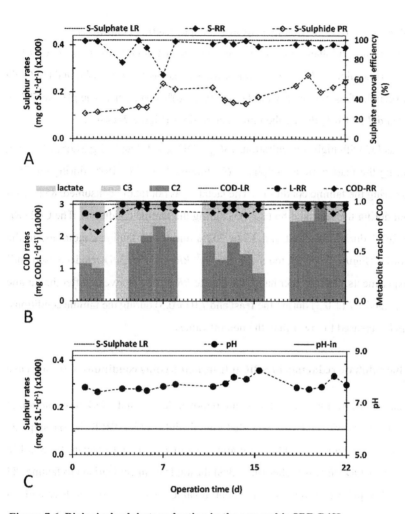

Figure 7-6. Biological sulphate reduction in the anaerobic SBR R1H

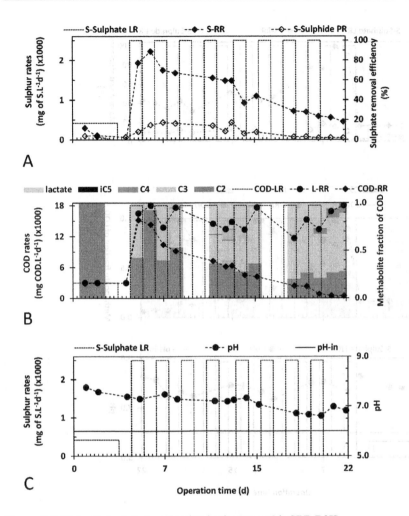

Figure 7-7. Biological sulphate reduction in the anaerobic SBR R2H

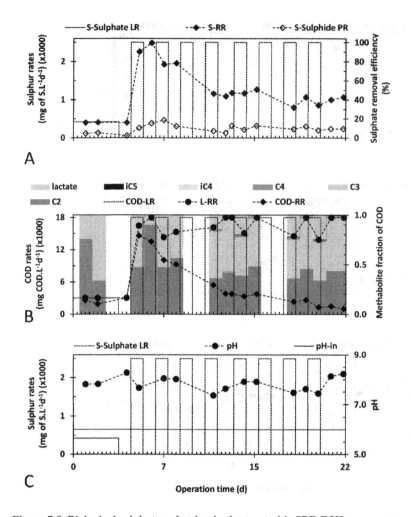

Figure 7-8. Biological sulphate reduction in the anaerobic SBR R3H

Table 7-1. Sulphate reduction at steady and transient feeding conditions

SBR	Lactate removal efficiency (%)	COD removal efficiency (%)	Sulphate removal efficiency (%)	Sulphide (mg.L^{-1})	pH
Control R1L					
Average	100	100	79±17	73±23	7.4±0.2
Max	100	100	98	105	7.7
Min	100	100	40	24	7.2
R2L Famine					
Average	100	92±6	90±11	331±168	7.5±0.1
Max	100	100	98	509	7.7
Min	100	85	63	42	7.4
R2L Feast					
Average	100	99±1	92±13	363±195	7.7±0.1
Max	100	100	99	583	7.9
Min	100	97	66	56	7.5
R3L Famine					
Average	99±4	85±8	86±11	379±179	7.4±0.1
Max	100	100	99	570	7.6
Min	88	73	67	79	7.2
R3L Feast					
Average	100	96±3	92±11	410±191	7.6±0.1
Max	100	100	99	544	7.7
Min	100	93	69	76	7.4

Continuation of Table 7-1

SBR	Lactate removal efficiency (%)	COD removal efficiency (%)	Sulphate removal efficiency (%)	Sulphide (mg.L^{-1})	pH
Control R1H					
Average	99±3	93±7	94±8	353±98	7.7±0.2
Max	100	100	100	542	8.3
Min	89	71	67	219	7.4
R2H Famine					
Average	86±14	35±35	36±29	315±287	7.0±0.3
Max	100	100	88	818	7.4
Min	65	2	14	89	6.6
R2H Feast					
Average	85±8	42±26	55±19	560±292	7.2±0.3
Max	98	84	76	867	7.5
Min	76	13	25	152	6.7
R3H Famine					
Average	92±10	33±34	55±28	440±185	7.8±0.3
Max	100	100	99	761	8.3
Min	77	6	29	117	7.4
R3H Feast					
Average	93±8	37±26	57±21	609±184	7.7±0.3
Max	100	81	90	919	8.0
Min	79	15	40	358	7.4

7.4   Discussion

7.4.1   Sulphate reduction process at transient feeding conditions in the SBR

This study showed that the sulphate reducing process is robust and resilience to transient feeding conditions and high sulphate concentration at no COD limiting. The COD:sulphate ratio was 2.4 along the experiments, thus, this ratio was never modified for avoiding disturbances different than the transient feeding conditions in the SBR. The COD:sulphate ratio is reported to be the most important parameter controlling the production of sulphide by means of dissimilatory sulphate reduction by SRB (Velasco et al., 2008). Additionally, according to the literature, a sulphate $RE > 90\%$ is guaranteed when lactate is used as electron donor at a COD:sulphate ratio of 2.4 (Torner-Morales and Buitrón, 2010).

The experiments R1L showed a sulphate RE of 79 ($\pm$ 17)%. The low concentration (1 g $COD_{Lactate}.L^{-1}$ and 0.4 g $SO_4^{2-}. L^{-1}$) supposed to avoid any drawback during the biological sulphate removal. Since lactate was completely removed (100% RE) as well as the COD (100% RE), other mechanisms for COD utilization were assumed, $e.g.$ methanogenesis, rather than sulphate reduction. For instance, methanogenic archaea are responsible of the utilization of hydrogen and acetate whereas sulphate reduction performing at a COD:sulphate ratio of 2.1 (O'Reilly and Colleran, 2006). But lactate had to be degraded first by other microorganism different that SRB, microorganism such as hydrolytic fermentative or homoacetogenic bacteria (Dar et $al.$, 2008; Wang et $al.$, 2008).

On the other hand, the R1H showed a sulphate RE of 94 ($\pm$ 8)%, using steady feeding conditions (at 6 g $COD_{Lactate}.L^{-1}$ and 2.5 g $SO_4^{2-}.L^{-1}$). Furthermore, using the same electron donor and acceptor concentrations, in the SBR R2L and R3L but during transient feeding conditions, the sulphate RE was $\geq$ 90%, unlike the sulphate RE of 86 ($\pm$ 11)% shown during famine conditions in R3L. Regardless the feast and famine conditions, lactate was consumed ($RE \geq 99\%$) in both SBR, R2L and R3L. Differently, the COD RE was lower during the famine periods, 92 ($\pm$ 6)% for R2L and 85 ($\pm$ 8)% for R3L, and higher during the feast periods, 99 ($\pm$ 1)% for R2L and 96 ($\pm$ 3)% for R3L. Most likely, the initial sulphate concentration during the feast periods is positively affecting the COD RE and, therefore, the COD RR (Figure 7-4B and Figure 7-5B) during the biological sulphate removal process. The biological COD and sulphate $RE \geq 90\%$ seems to be influenced by the initial sulphate concentrations (2.5 g $SO_4^{2-}.L^{-1}$) tested in the SBR R1H, R2L and R3L during the steady feeding and feast periods, respectively. Pure SRB cultures had an optimal specific growth rate at the initial sulphate concentration of 2.5 g $SO_4^{2-}.L^{-1}$ (Al-Zuhair et $al.$, 2008).

Moreover, the pH ranged from 7.2 to 8.3 during the biological sulphate removal in the SBR R1H, R2L and R3L, regardless the steady or transient feeding conditions wherein the influent pH was 6.0. Using lactate and propionate as electron donors, the biological sulphate reduction produces carbonate (Eq. 7-5 to Eq. 7-8). The buffering capacity of the produced carbonate is the reason of the pH values > 7.2 in the SBR (R1H, R2L and R3L). Other electron donors can be used by SRB apart from lactate and propionate, as reviewed in the literature (Liamleam and Annachhatre, 2007).

$$2 \text{ Lactate}^- + SO_4^{2-} \rightarrow 2 \text{ Acetate}^- + HS^- + 2HCO_3^- + H^+ \qquad \text{Eq. 7-5}$$

$$2\text{Lactate}^- + 3SO_4^{2-} \rightarrow 6 \text{ HCO}_3^- + HS^- + H^+ \qquad \text{Eq. 7-6}$$

$$\text{Propionate}^- + 0.75 \text{ } SO_4^{2-} \rightarrow \text{Acetate}^- + HCO_3^- + 0.75 \text{ } HS^- + 0.25 \text{ } H^+ \qquad \text{Eq. 7-7}$$

$$\text{Propionate}^- + 1.75 \text{ } SO_4^{2-} \rightarrow 3 \text{ HCO}_3^- + 1.75 \text{ } HS^- + 0.25 \text{ } H^+ \qquad \text{Eq. 7-8}$$

During transient feeding conditions, the biological sulphate reduction is more affected using higher electron donor and acceptor concentrations (36 g $COD_{Lactate}.L^{-1}$ and 15 g $SO_4^{2-}.L^{-1}$). The sulphate RE was reduced to < 20% in the SBR R2H and maintained at slightly above 40% in the SBR R3H at the end of the operation time (Figure 7-7A and Figure 7-8A). Therefore, the final pH decreased to < 7.0 in R2H and increased to > 8.0 in R3H. These pH changes are associated to sulphate removal (Eq. 7-5 to Eq. 7-8). Nevertheless, the average pH was preserved above 7.0 in the experiments using NH_4^+ (7.0-7.2 for R2H) and without NH_4^+ (7.7-7.8 for R3H) despite feast or famine feeding conditions during the operation time.

The COD RE dropped to reach almost zero values in the SBR R2H (~ 2%) as well as in R3H (~ 6%) at the end of the reactor operation. The average lactate RE (92-93%) was higher in the R3H, regardless the feast famine conditions, when compared to R2H (lactate RE = 85-86%). The low average COD RE ($\leq$ 42%) and the relative high lactate RE ($\geq$ 85%) suggest that the biological sulphate removal was carried out using, mainly, lactate as the electron donor. Sulphide inhibition was discarded (818-919 mg.L^{-1}), some bioreactors performing sulphate reduction at 1,215 mg.L^{-1} sulphide concentrations have not been inhibited (Celis-García et al., 2007). However, the potential toxicity of sulphide increases simultaneously to the increments of the HRT (Kaksonen et al., 2004).

The biological sulphate reduction in the SBR R2L and R3L has overcome the transient feeding conditions and managed to re-establish the condition similar those in the control SBR R1H. In this way and according to the literature (Kitano, 2007), the resilience and robustness of sulphate reduction is demonstrated in the SBR. Moreover, the quantity of enzymes capable to perform a desired reaction is related to parameters of robustness like the time of adaptation (Alon et al., 1999). The sulphate removal can overcome low RE efficiencies when the regime of operation is changed, e.g. from a continuous to a continuous with biomass recirculation operation (Boshoff et al., 2004). The dynamic of the bioreactors imposed by a proper HRT can concentrate the desired amount of enzymes by the SRB to perform sulphate reduction at high rates.

7.5 Conclusions

This research has demonstrated that sulphate reduction is vulnerable (in R1L) when low electron donor and acceptor concentrations (1 g $COD_{Lactate}.L^{-1}$ and 0.4 g $SO_4^{2-}.L^{-1}$) are used, resulting in a sulphate RE of 79 ($\pm$ 17)%. On the other hand, the biological sulphate removal is robust to transient feeding conditions in SBR R2L and R3L. The sulphate RE was very similar during the feast (92%) and famine (86-90%) in both SBR, R2L and R3L, when compared to the control reactor R1H (94 $\pm$ 8%). The sulphate RE was severally affected during transient feeding conditions at 36 g $COD_{Lactate}.L^{-1}$ and 15 g $SO_4^{2-}.L^{-1}$. The sulphate removal dropped to < 20 % in the SBR R2H and was maintained at slightly above 40 % in the SBR R3H.

7.6 References

Al-Zuhair, S., El-Naas, M.H., Al-Hassani, H., 2008. Sulfate inhibition effect on sulfate reducing bacteria. J. Biochem. Technol. 1, 39–44.

Alon, U., Surette, M.G., Barkai, N., Leibler, S., 1999. Robustness in bacterial chemotaxis. Nature 397, 168–171. doi:10.1038/16483

APHA, 1999. Standard Methods for the Examination of Water and Wastewater, 20th ed. Washington, USA.

Barkai, N., Leibler, S., 1997. Robustness in simple biochemical networks. Nature 387, 913–917. doi:10.1038/43199

Bertolino, S.M., Rodrigues, I.C.B., Guerra-Sá, R., Aquino, S.F., Leão, V.A., 2012. Implications of volatile fatty acid profile on the metabolic pathway during continuous sulfate reduction. J. Environ. Manage. 103, 15–23. doi:10.1016/j.jenvman.2012.02.022

Boshoff, G., Duncan, J., Rose, P., 2004. Tannery effluent as a carbon source for biological sulphate reduction. Water Res. 38, 2651–2658. doi:10.1016/j.watres.2004.03.030

Celis-García, L.B., Razo-Flores, E., Monroy, O., 2007. Performance of a down-flow fluidized bed reactor under sulfate reduction conditions using volatile fatty acids as electron donors. Biotechnol. Bioeng. 97, 771–779. doi:10.1002/bit.21288

Dar, S.A., Kleerebezem, R., Stams, A.J.M., Kuenen, J.G., Muyzer, G., 2008. Competition and coexistence of sulfate-reducing bacteria, acetogens and methanogens in a lab-scale anaerobic bioreactor as affected by changing substrate to sulfate ratio. Appl. Microbiol. Biotechnol. 78, 1045–1055. doi:10.1007/s00253-008-1391-8

Dries, J., De Smul, A., Goethals, L., Grootaerd, H., Verstraete, W., 1998. High rate biological treatment of sulfate-rich wastewater in an acetate-fed EGSB reactor. Biodegradation 9, 103–111. doi:10.1023/A:1008334219332

Kaksonen, A.H., Franzmann, P.D., Puhakka, J.A., 2004. Effects of hydraulic retention time and sulfide toxicity on ethanol and acetate oxidation in sulfate-reducing metal-precipitating fluidized-bed reactor. Biotechnol. Bioeng. 86, 332–343. doi:10.1002/bit.20061

Kitano, H., 2007. Towards a theory of biological robustness. Mol. Syst. Biol. 3, 1–7. doi:10.1038/msb4100179

Liamleam, W., Annachhatre, A.P., 2007. Electron donors for biological sulfate reduction. Biotechnol. Adv. 25, 452–463. doi:10.1016/j.biotechadv.2007.05.002

Mapanda, F., Nyamadzawo, G., Nyamangara, J., Wuta, M., 2007. Effects of discharging acid-mine drainage into evaporation ponds lined with clay on chemical quality of the surrounding soil and water. Phys. Chem. Earth, Parts A/B/C 32, 1366–1375. doi:10.1016/j.pce.2007.07.041

Mottet, A., Habouzit, F., Steyer, J.P., 2014. Anaerobic digestion of marine microalgae in different salinity levels. Bioresour. Technol. 158, 300–306. doi:10.1016/j.biortech.2014.02.055

O'Reilly, C., Colleran, E., 2006. Effect of influent COD/SO$_4^{2-}$ ratios on mesophilic anaerobic reactor biomass populations: physico-chemical and microbiological properties. FEMS Microbiol. Ecol. 56, 141–153. doi:10.1111/j.1574-6941.2006.00066.x

Okabe, S., Nielsen, P.H., Charcklis, W.G., 1992. Factors affecting microbial sulfate reduction by Desulfovibrio desulfuricans in continuous culture: limiting nutrients and sulfide concentration. Biotechnol. Bioeng. 40, 725–734. doi:10.1002/bit.260400612

Plugge, C.M., Zhang, W., Scholten, J.C.M., Stams, A.J.M., 2011. Metabolic flexibility of sulfate-reducing bacteria. Front. Microbiol. 2, 1–8. doi:10.3389/fmicb.2011.00081

Quéméneur, M., Hamelin, J., Latrille, E., Steyer, J.-P., Trably, E., 2011. Functional versus phylogenetic fingerprint analyses for monitoring hydrogen-producing bacterial populations in dark fermentation cultures. Int. J. Hydrogen Energy 36, 3870–3879. doi:10.1016/j.ijhydene.2010.12.100

Reyes-Alvarado, L.C., Okpalanze, N.N., Kankanala, D., Rene, E.R., Esposito, G., Lens, P.N.L., 2017. Forecasting the effect of feast and famine conditions on biological sulphate reduction in an anaerobic inverse fluidized bed reactor using artificial neural networks. Process Biochem. 55, 146–161. doi:10.1016/j.procbio.2017.01.021

Sipma, J., Osuna, M.B., Lettinga, G., Stams, A.J.M., Lens, P.N.L., 2007. Effect of hydraulic retention time on sulfate reduction in a carbon monoxide fed thermophilic gas lift reactor. Water Res. 41, 1995–2003. doi:10.1016/j.watres.2007.01.030

Torner-Morales, F.J., Buitrón, G., 2010. Kinetic characterization and modeling simplification of an anaerobic sulfate reducing batch process. J. Chem. Technol. Biotechnol. 85, 453–459. doi:10.1002/jctb.2310

Velasco, A., Ramírez, M., Volke-Sepúlveda, T., González-Sánchez, A., Revah, S., 2008. Evaluation of feed COD/sulfate ratio as a control criterion for the biological hydrogen sulfide production and lead precipitation. J. Hazard. Mater. 151, 407–413. doi:10.1016/j.jhazmat.2007.06.004

Villa-Gomez, D.K., Papirio, S., van Hullebusch, E.D., Farges, F., Nikitenko, S., Kramer, H., Lens, P.N.L., 2012. Influence of sulfide concentration and macronutrients on the characteristics of metal precipitates relevant to metal recovery in bioreactors. Bioresour. Technol. 110, 26–34. doi:10.1016/j.biortech.2012.01.041

Wang, A., Ren, N., Wang, X., Lee, D., 2008. Enhanced sulfate reduction with acidogenic sulfate-reducing bacteria. J. Hazard. Mater. 154, 1060–1065. doi:10.1016/j.jhazmat.2007.11.022

Whiteley, C.G., Lee, D.-J., 2006. Enzyme technology and biological remediation. Enzyme Microb. Technol. 38, 291–316. doi:10.1016/j.enzmictec.2005.10.010

Chapter 8

Carbohydrate based polymeric materials as slow release electron donors for sulphate removal from wastewater

A modified version of this chapter was published as:
Reyes-Alvarado L.C., Okpalanzea N.N., Rene E.R., Rustrian E., Houbron E., Esposito G., Lens P.N.L., (2017). Carbohydrate based polymeric materials as slow release electron donors for sulphate removal from wastewater. J Environ Manage. doi: 10.1016/j.jenvman.2017.05.074

Abstract

Many industrial sulphate rich wastewaters are deficient in electron donors to achieve complete sulphate removal. Therefore, pure and expensive chemicals are supplied externally. In this study, carbohydrate based polymers (CBP) as potato (2 and 5 mm^3), filter paper (2 and 5 mm^2) and crab shell (2 and 4 mm Ø) were tested as slow release electron donors (SRED) for biological sulphate reduction at 30 °C and initial pH of 7.0. Using the CBP as SRED, sulphate reduction was carried out at different rates: filter paper 0.065-0.050 > potato 0.022-0.034 > crab shell 0.006-0.009 mg SO$_4^{2-}$.mg VSS^{-1}d^{-1}. These were also affected by the hydrolysis-fermentation rates: potato 0.087-0.070 > filter paper 0.039-0.047 > crab shell 0.011-0.028 mg CODs.mg VSS^{-1}d^{-1}, respectively. Additionally, the sulphate removal efficiencies using filter paper (cellulose, > 98%), potato (starch, > 82%) and crab shell (chitin, > 32%) were achieved only when using CBP as SRED and in the absence of other easily available electron donors. This study showed that the natural characteristics of the CBP limited the hydrolysis-fermentation step and, therefore, the sulphate reduction rates.

Keywords: Sulphate reducing bacteria, hydrolytic-fermentative bacteria, carbohydrate based polymers, slow release electron donors

8.1 Introduction

The water cycle is highly affected by the growing population that demands for goods and facilities. Sulphur species are present in the water cycle by anthropogenic activities; for example, food and electronics production, mineral extraction, pulp and paper and petrochemical industry. Sulphur species are also discharged as sulphate in vinasses (Robles-González et al., 2012) from alcoholic beverage production or in acid mine drainage (AMD) (Nieto et al., 2007), dimethyl sulphoxide in semi-conductor production (Park et al., 2001), sulphite from pulping (Pokhrel and Viraraghavan, 2004) and sulphide from petrochemical industries. Some of these water streams are characterized by the presence of a complex mixture of heavy and toxic metals, e.g. from AMD (Borrego et al., 2012; Monterroso and Macías, 1998) and the electronic industry (Rengaraj et al., 2003).

Another important characteristic of many industrial wastewaters is the very low content of chemical oxygen demand (COD), which is in some cases lower than 100 mg.L^{-1} (Bai et al., 2013; Deng and Lin, 2013) and insufficient to remove sulphate by sulphate reducing bacteria (SRB). Hence, the addition of an external carbon source is required, which in most cases increases the treatment cost due to the use of expensive sources, e.g. lactate and formate, or they might require special safety installations, e.g. when hydrogen is used (Liamleam and Annachhatre, 2007). The operating costs can be reduced when organic wastes are used as electron donors, which also make the treatment process more sustainable. Cheese whey (Martins et al., 2009), molasses (Wang et al., 2008), plant hydrolyzates (Lakaniemi et al., 2010), horse manure and vegetable compost (Castillo et al., 2012) are some examples of wastes that have been used as alternate electron donors for sulphate removal.

Hydrolysis-fermentation is the rate limiting step in anaerobic digestion of organic solid wastes (Houbron et al., 2008). This low rate of organic matter decomposition provides slow release electron donor (SRED): low molecular weight compounds or soluble COD (CODs) are provided for to the next trophic levels, including sulphate reduction. Hence, carbohydrate based polymers (CBP), such as starch from potato, cellulose from filter paper and chitin from crab shell cannot be directly used as electron donors by SRB, but the hydrolytic-fermentative bacteria can use them to produce a mixture of volatile fatty acids (VFA), alcohols, ketones, and other low molecular weight compounds that can then be used by the SRB to perform sulphate reduction.

Different carbohydrates (starch, cellulose and chitin) have different degrees of complexity in their structure, e.g. crystallinity, and are thus hydrolyzed and fermented at different rates

(Jeihanipour et al., 2011; Labatut et al., 2011). Consequently, the sulphate reduction rate will depend on the CBP supplied as SRED to a bioreactor or permeable barrier. The objective of this research was, therefore, to study the sulphate removal from synthetic wastewater using the COD_S released during the hydrolysis-fermentation of different types of CBP by hydrolytic-fermentative bacteria. Also, the COD_S released from the CBP to the synthetic wastewater without inoculum as well as with inoculum were investigated. In this way, the COD_S released naturally from the CBP, the COD_S released by the hydrolysis-fermentation of the CBP and the COD_S released and used for sulphate reduction using CBP as SRED could be differentiated. Furthermore, the biochemical activities were determined in batch bioreactors.

8.2   Material and methods

8.2.1   Inoculum

The inoculum was obtained from an anaerobic reactor treating activated sludge at the municipal wastewater treatment plant, located at Harnaschpolder (The Netherlands). The seed liquid contained a total suspended solid (TSS) concentration of 30.8 ($\pm$ 2.1) g $TSS.L^{-1}$ and volatile suspended solids (VSS) concentration of 20.4 ($\pm$ 1.5) g $VSS.L^{-1}$. TSS and VSS were concentrated (5000×g, 10 min and 4°C) and the solid phase was used as inoculum in the batch bioreactors to obtain a constant initial concentration of 2 g $VSS.L^{-1}$.

8.2.2   CBP as electron donors

The following CBP were used as SRED at two different particle sizes: starch supplied as potato (2 and 5 mm^3), cellulose as filter paper (2 and 5 mm^2) and chitin as crab shell (2 and 4 mm diameter, Ø). CBP samples of potato and filter paper were cut with a kitchen knife and sizes were measured with a ruler. The crab shell was broken with a hammer, properly cleaned with acetone prior to its use, and further sized down with a kitchen blender, the size selection was made with a sieve (mesh 5-6).

The COD was analysed for each CBP, this COD was named recalcitrant COD (COD_R) because it needs to be chemically or biologically hydrolysed prior to the COD analysis (Vaccari et al., 2005), e.g. lactate is soluble and gives COD_S. The organic solid samples (1 g of CBP) were hydrolysed with a mixture of sulphuric acid (5 mL) and MiliQ water (5 mL) at 70 °C for 3 h (Lenihan et al., 2011) and the COD_S was determined as described in section 8.2.7.

8.2.3 Synthetic wastewater composition

The composition of the synthetic wastewater used in this study was as follows (mg.L^{-1}): NH$_4$Cl (300), MgCl$_2$•6H$_2$O (120), KH$_2$PO$_4$ (200), KCl (250), CaCl$_2$•2H$_2$O (15), yeast extract (20) and 0.5 mL of a mixture of micronutrients. The trace elements had the following composition (mg.L^{-1}): FeCl$_2$•4H$_2$O (1500), MnCl$_2$•4H$_2$O (100), EDTA (500), H$_3$BO$_3$ (62), ZnCl$_2$ (70), NaMoO$_4$•2H$_2$O (36), AlCl$_3$•6H$_2$O (40), NiCl$_3$•6H$_2$O (24), CoCl$_2$•6H$_2$O (70), CuCl$_2$•2H$_2$O (20) and HCl 36 % (1 mL) (Villa-Gomez et al., 2011). Sodium lactate (was solely used for sulphate reducing activity tests of the biomass) and sodium sulphate were also used as, respectively, the electron donor and acceptor when needed. All reagents used in this study were of analytical grade.

8.2.4 Sulphate reducing and methanogenic activity test of anaerobic sludge

Methanogenic activity of the biomass was determined as described by Angelidaki et al., (2009) and glucose was used as the electron donor. The sulphate reducing activity of the biomass was determined as described by Villa-Gomez et al., (2011) using sulphate and lactate (as CODs) as electron acceptor and donor, respectively, at a constant ratio of 1:1. The experiments were performed in batch (serum bottles of 500 mL), filled up to 0.3 L of mineral media and 0.2 L of headspace, covered with an airtight rubber stopper and done in triplicates. The initial pH was adjusted to 7.0. Each batch bioreactor was flushed with nitrogen and the initial pressure in the serum bottle was kept constant at 1 bar. The batch bioreactors were maintained at 30 °C and agitated at 160 rpm on an orbital shaker (New Brunswick Scientific Innova 2100 platform shaker, Eppendorf, USA). Methanogenic activity was evaluated as a response of the pressure increment in the head space. The pressure increment of each batch incubation was recorded with a manometer (LEO 1 digital manometer, Winterthur, Switzerland) and used to calculate the volume of biogas produced considering a theoretical biogas composition (CH$_4$/CO$_2$ = 70/30% v/v, Yentekakis, 2006) and following the calculation explained by de Lemos Chernicharo (2007). Biogas was represented as mg COD-CH$_4$.L^{-1} in the plots, however, the composition was not analysed, but only simulated.

8.2.5 SRED experiments

The SRED experiments were done in triplicate and carried out in batch, 500 mL serum bottles fitted with airtight rubber stoppers, as described above. CBP (1.02 ± 0.01 g potato, filter paper or crab shell) were added individually as the sole source of electron donor and under no circumstance lactate was added. The CBP were investigated as follows: Test 1 for CODs release in synthetic wastewater under anaerobic conditions without inoculum and under non-sterile conditions. Test 2 for CODs release with inoculum (hydrolysis-fermentation), and test 3, for CODs release and sulphate reduction (by addition of sulphate, 760 mg.L^{-1} to the synthetic wastewater) with inoculum.

8.2.6 Estimation of volumetric and specific rates

The volumetric rates (V_r), specific rates (S_r) and sulphate removal efficiencies (SRE) were evaluated using the following equations:

$$V_r = \frac{(y_0 - y_1)}{(t_1 - t_0)}$$ Eq. 8-1

$$S_r = \frac{V_r}{[VSS]}$$ Eq. 8-2

$$SRE = \left(\frac{(SO_{4\,initial}^{2-} - SO_{4\,final}^{2-})}{SO_{4\,initial}^{2-}}\right) \times 100$$ Eq. 8-3

Where, y_0 is the concentration of any parameter at the beginning of the experiment (t_0) and y_1 is the concentration of the same parameter after the time of reaction (t_1) and [VSS] is the concentration of volatile suspended solids.

8.2.7 Analysis

The pH was measured with a sulphide resistant electrode (Prosense, Oosterhout, The Netherlands). Sulphate was analyzed by ion chromatography, as described elsewhere (Villa-Gomez et al., 2011). The total solids (TS) concentration was estimated after the removal of moisture from the solid sample, at 105 °C, using an oven. The ash content was determined after burning the sample in a muffle furnace at 550 °C. The volatile solids (VS) were obtained by the extraction of the ash content from TS. The CODs (determined by the close reflux colorimetric method), TS, moisture content, VS, ash content, VSS, TSS and total dissolved sulphide (S^{2-} or sulphide) were measured according to the procedures outlined in the Standard Methods (APHA, 1999).

8.3 Results

8.3.1 Sulphate reduction and methanogenic activity of the anaerobic inoculum

During the methanogenic activity test, biogas production shows a lag phase of 11 d, an exponential gas production from 11 to 46 d and a stationary phase between 46-80 d (Figure

8-1A). The *Vr* and *Sr* activities of the sludge were estimated to be 70.5 mg COD-CH₄.L⁻¹.d⁻¹ and 0.035 mg COD-CH4.mg $VSS^{-1}.d^{-1}$, respectively.

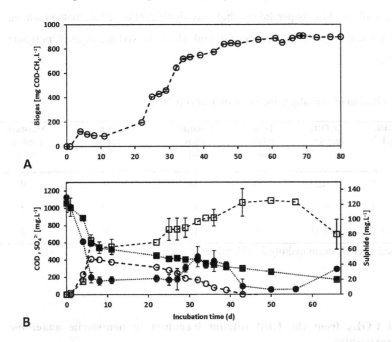

Figure 8-1. Biochemical activities shown by the inoculum
A) Biogas production activity (hydrolytic-fermentative bacteria, homoacetogens and methanogens), biogas represented as COD-CH₄ (○). B) sulphate reduction by SRB. Lactate as CODs (●), biogas represented as COD-CH₄ (○), sulphate (■) and sulphide (□)

During the first 10 h of the sulphate reduction activity test, sulphate and lactate (as CODs) were consumed simultaneously, thus higher rates were observed during this period (Figure 8-1B). The *Vr* and *Sr* activities were estimated to be: 144 mg $SO_4^{2-}.L^{-1}.d^{-1}$ and 0.072 mg $SO_4^{2-}.mg$ $VSS^{-1}.d^{-1}$, respectively. The CODs consumption occurred simultaneously, but at different rates: 207 mg $CODs.L^{-1}.d^{-1}$ or 0.1035 mg CODs.mg $VSS^{-1}.d^{-1}$. As a consequence of the sulphate reduction, sulphide was produced at a rate of 27 mg $S^{2-}.L^{-1}.d^{-1}$ or 0.014 mg $S^{2-}.mg$ $VSS^{-1}.d^{-1}$. During the first 6 d, an exponential sulphide production was observed, followed by a decrease in the rate. After 20 d, the rate of sulphide production increased again.

8.3.2   CBP characteristics
The CODᴿ values of the CBP materials were 1.5, 0.38, and 0.53 g.g VS^{-1} for potato, filter paper and crab shell, respectively (Table 8-1). However, this CODᴿ was only measured after a chemical hydrolysis of the respective materials. The moisture content, inorganic matter and VS

content for each of the CBP tested were also different. Potato was the CBP with the highest moisture content (80%), whereas filter paper (3%) and crab shell (2%) had a much lower moisture content. In addition, filter paper did not show any significant inorganic matter content, but the amount of VS was high (97%). In contrast, crab shell showed the highest inorganic matter content (79%).

Table 8-1. Characterization of carbohydrate based polymers (CBP)

CBP	$CODs$* (g.L^{-1})	COD_R (g.g VS^{-1})	Total solids (%)	Volatile solids (%)	Inorganic solids (%)	Moisture content (%)
Potato	27.1	1.5	20	18	2	80
Filter paper	37.3	0.38	97	97	0	3
Crab shell	10.2	0.53	98	19	79	2

Note: *$CODs$ analysed after chemical hydrolysis with sulphuric acid

8.3.3 Release of $CODs$ from the CBP without inoculum in non-sterile anaerobic synthetic wastewater

The self release of $CODs$ without inoculum in non-sterile synthetic wastewater was studied to differentiate the $CODs$ that can be released from the CBP due to the activity of endogenous hydrolytic-fermentative bacteria present in the inoculum. During these experiments (Figure 8-2), COD in the form of low molecular weight compounds ($CODs$) diffused out of the CBP, which are formed as intermediates during the synthesis of the organic solids. When potato was placed in the anaerobic synthetic wastewater (neither sulphate nor inoculum were added), $CODs$ was released from the potato and it showed a peak at 11 d for both particle sizes: 503 and 405 mg $CODs$.L^{-1} for particles of 2 and 5 mm^3, respectively. The profile of $CODs$ released from the filter paper showed a lag phase for both particle sizes investigated. After 21 d, 231 mg $CODs$.L^{-1} was released for the larger particle size (5 mm^2) and after 45 d, 214 mg $CODs$.L^{-1} was also released for the smaller particle size (2 mm^2). Only 89 mg $CODs$.L^{-1} was released when the smaller particle size (2 mm Ø) of crab shell were used. The $CODs$ released without inoculum to the synthetic wastewater did not reach the potential COD_R found in each CBP.

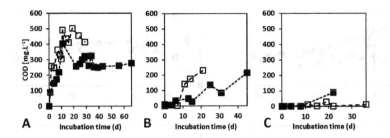

Figure 8-2. CODs released in non-sterile anaerobic synthetic wastewater without inoculum
The smaller particle size is represented with (■) and the larger with (□). A) potato 2 mm³ size, potato 5 mm³ size, B) filter paper 2 mm², filter paper 5 mm², C) crab shell 2 mm Ø and crab shell 4 mm Ø

8.3.4 Release of CODs from the CBP in the presence of inoculum

The CBP used as SRED were placed under anaerobic conditions in a mixture of synthetic wastewater and inoculum. In the case of potato, 2 and 5 mm³ particle sizes, CODs increased to the highest level (605-789 mg CODs.L⁻¹) within the first 10 d, followed by its consumption (Figure 8-3A-B). Experiments with filter paper (Figure 8-3C-D) showed a constant increase in the release of CODs, 1123 and 1348 mg CODs.L⁻¹ were the highest CODs concentrations observed for the particle size of 2 and 5 mm², respectively. Crab sell showed the lowest CODs release of 181 and 364 mg CODs.L⁻¹ for particles with a diameter of 2 and 4 mm, respectively (Figure 8-3E-F). CODs release was performed at different rates for the CBP and a decrease in pH was observed for potato (6.3-6.2) and filter paper (5.8-5.1) for both particle sizes used, from an initial pH of 7.0. On the other hand, the pH of crab shell incubations increased to 7.1 and 7.2 for the particles with 2 and 4 mm of diameter, respectively.

Based on the calculations (Table 8-2), potato was an easy available carbon source for hydrolysis-fermentation (0.087-0.07 mg CODs.mg VSS⁻¹d⁻¹), followed by filter paper (0.039-0.047 mg CODs.mg VSS⁻¹d⁻¹), whereas crab shell was the CBP with the lowest CODs availability (0.011-0.028 mg CODs.mg VSS⁻¹d⁻¹).

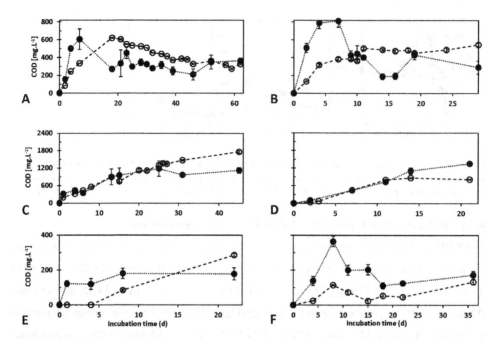

Figure 8-3. COD$_S$ release in synthetic wastewater with inoculum

The COD$_S$ in the liquid phase (●) and cumulative biogas in the head space represented as COD-CH$_4$ (○). A) potato 2 mm^3 size, B) potato 5 mm^3 size, C) filter paper 2 mm^2, D) filter paper 5 mm^2, E) crab shell 2 mm Ø and F) crab shell 4 mm Ø

Table 8-2. Kinetic data calculated from experiments related to COD$_S$ released from the CBP by the influence of inoculum in synthetic wastewater

		CBP					
	Parameter	Potato		Filter paper		Crab shell	
	Particle size	2 mm^3	5 mm^3	2 mm^2	5 mm^2	2 mm Ø	4 mm Ø
COD	Volumetric rate (mg COD$_S$.L^{-1}d^{-1})	173	139	78	94	22	56
	Specific rate (mg COD$_S$.mg VSS^{-1}d^{-1})	0.087	0.070	0.039	0.047	0.011	0.028
pH	Initial	7.0	7.0	7.0	7.0	7.0	7.0
	Final	6.3	6.2	5.8	5.1	7.1	7.2

8.3.5 Sulphate reduction during the release of CODs from the CBP in the presence of inoculum

When potato cubes were used, they were hydrolysed and fermented within the first 4 d, followed by CODs depletion. The hydrolysis-fermentation rates observed were 0.033 and 0.026 mg CODs.mg VSS^{-1}d^{-1}, for 2 and 5 mm^3 particle sizes, respectively. Sulphate reduction showed activity according to the CODs profiles. The observed specific sulphate removal rates were 0.022 and 0.034 mg SO$_4^{2-}$.mg VSS^{-1}d^{-1} for the particle sizes of 2 and 5 mm^3, respectively. Despite the difference in the size of the CBP, both sulphate reduction profiles showed the same decreasing trends. The maximum sulphide concentrations were 77 and 35 mg S^{2-}.L^{-1} for potato cubes of 2 and 5 mm^3 particle sizes, respectively (Figure 8-4A-B).

During the experiments with filter paper, CODs levels were very low ($<$ 300 mg CODs.L^{-1}) and they only increased after the sulphate was consumed (Figure 8-4C-D). Irrespective of the filter paper size, sulphate was reduced in less than 13 d. Specific sulphate reduction rates were the highest at 0.065 and 0.05 mg SO$_4^{2-}$.mg VSS^{-1}d^{-1} for 2 and 5 mm^2 particle sizes, respectively. The sulphide concentrations were also the highest in these experiments: 81 and 106 mg S^{2-}.L^{-1} for 2 and 5 mm^2 particle sizes, respectively. Concerning the profiles of CODs with crab shell, their values were the lowest at approximately 200 mg CODs.L^{-1}. The specific sulphate reduction rates of 0.006 and 0.009 mg SO$_4^{2-}$.mg VSS^{-1}d^{-1} were observed for 2 and 4 mm Ø particle sizes, respectively.

The media pH was adjusted to 7.0 in all the batch experiments. After the sulphate reduction incubation, when potato was used, the pH decreased to 6.8 and 6.7 for 2 and 5 mm^3 particle sizes, respectively. For experiments with filter paper, the final pH was 6.2 for both particle sizes investigated and for crab shell, the pH increased to 7.3 and 7.5 for particles with a diameter of, respectively, 2 and 4 mm.

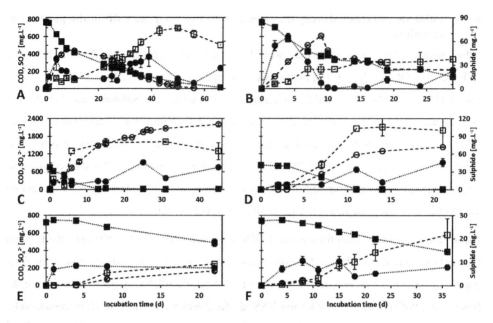

Figure 8-4. Sulphate reduction profiles using CBP as SRED in synthetic wastewater
The COD$_S$ in the liquid phase (●), biogas in the head space represented as COD-CH$_4$ (○), sulphate (■) and total dissolved sulphide (□). A) potato 2 mm^3 size, B) potato 5 mm^3 size, C) filter paper 2 mm^2, D) filter paper 5 mm^2, E) crab shell 2 mm Ø and F) crab shell 4 mm Ø

8.4   Discussion

8.4.1   Use of carbohydrate based polymeric materials as slow release electron donors

This work shows that CBP can be used as SRED using anaerobic sludge as the source of SRB (Figure 8-4). Hereby, the CBP are mainly composed of starch, cellulose or chitin, respectively, but other functional compounds and precursors of these carbohydrates are present in the solid matrix as well (Pedreschi *et al.*, 2009). Most likely, glucose and soluble proteins were predominantly responsible for the COD$_S$ released in the case of potato cube experiments (Pedreschi *et al.*, 2009). In the paper production, the pulping process produces fines or low molecular weight saccharides which are responsible for the released COD$_S$ (Chevalier-Billosta *et al.*, 2006). Chitin is poorly soluble in aqueous environments and this characteristic is also related to the semi crystalline conformation of the carbohydrate (Pillai *et al.*, 2009). These carbohydrate compounds diffuse out of the CBP material when they are contacted with anaerobic synthetic wastewater (Figure 8-2), and can subsequently be used as electron donor for the SRB. The COD$_S$ released was in the following order: potato > filter paper > crab shell.

Table 8-3. Kinetic data and efficiencies calculated from experiments related to sulphate reduction using CBP as SRED in synthetic wastewater

	Parameter	CBP					
		Potato		Filter paper		Crab shell	
	Particle size	2 mm^3	5 mm^3	2 mm^2	5 mm^2	2 mm Ø	4 mm Ø
COD	Volumetric rate (mg COD$_S$.L^{-1}d^{-1})	67	52	15	83	13	23
	Specific rate (mg COD$_S$.mg VSS^{-1}d^{-1})	0.033	0.026	0.007	0.042	0.006	0.011
Sulphate	Volumetric rate (mg SO$_4^{2-}$.L^{-1}d^{-1})	-43	-68	-129	-101	-13	-18
	Specific rate (mg SO$_4^{2-}$.mg VSS^{-1}d^{-1})	-0.022	-0.034	-0.065	-0.050	-0.006	-0.009
	Initial (mg.L^{-1})	766	764	752	823	721	738
	Final (mg.L^{-1})	14	139	17	9	488	386
	Removed (mg.L^{-1})	752	625	735	815	233	352
	S-SO$_4^{2-}$ Removed (mg.L^{-1})	250	208	245	271	78	117
	Removal Efficiency (%)	98	82	98	99	32	48
Sulphide	Volumetric rate (mg S^{2-}.L^{-1}d^{-1})	4.6	5.4	17.9	15.4	1.3	1.3
	Specific rate (mg S^{2-}.mg VSS^{-1}d^{-1})	0.0023	0.0027	0.0089	0.0077	0.0006	0.0007
	Highest concentration (mg S^{2-}.L^{-1})	77	35	81	106	9	22
pH	Initial	7.0	7.0	7.0	7.0	7.0	7.0
	Final	6.8	6.7	6.2	6.2	7.3	7.5

8.4.2 Biological sulphate reduction using CBP as SRED

Under the conditions tested, the highest sulphate reduction efficiencies and rates were expected to occur using potato, mainly by considering the COD$_S$ released in anaerobic synthetic wastewater in the absence of inoculum (potato > filter paper > crab shell, Figure 8-2) and the rate of COD$_S$ released (mg COD$_S$.mg VSS^{-1}d^{-1}) in anaerobic synthetic wastewater by the influence of the inoculum (potato > filter paper > crab shells, Table 8-2). However, the filter paper offered higher sulphate removals efficiencies (> 98%) compared to potato (> 82%) and crab shell (> 32%). This suggests that under the hydrolysis-fermentation conditions and in the absence of sulphate, α-glycosidase can perform optimally the hydrolysis of starch (present in potato as CBP) when compared to β-glycosidase during the lower hydrolysis rate of cellulose (present in paper as CBP). Furthermore, the activity of the α-glycosidase enzyme is probably inhibited under sulphate reducing conditions. On the other hand, under sulphate reducing conditions, the β-glycosidase performs optimally. The proportion of cellulose degrading microorganisms is much more abundant in comparison to other groups (like SRB or methanogenic archeae) under sulphate reducing conditions and the use of cellulosic biowaste as carbon source (Pruden et al., 2007).

The CODs released in anaerobic synthetic wastewater in the presence of inoculum (due to the process of hydrolysis-fermentation) decreased the pH: 7.1-7.2 for 2 and 4 mm Ø crab shells > 6.3-6.2 for 2 and 5 mm^3 potato > 5.8-5.1 for 2 and 5 mm^2 filter paper (Table 8-2); such pH changes suggest the production of VFA's. During the biological sulphate reduction, the pH changes were not in the same order of magnitude using CBP as SRED: 7.3-7.5 for 2 and 4 mm Ø crab shells > 6.8-6.7 for 2 and 5 mm^3 potato > 6.2 for 2 and 5 mm^2 filter paper (Table 8-3), the pH were less acid for potato cubes and filter paper in contrast to those observed in the hydrolysis-fermentation incubations. These pH differences between the hydrolysis-fermentation and sulphate reduction experiments, both using the CBP as SRED, indicate the utilization of VFA by SRB for sulphate reduction and production of CO_2. Nevertheless, the pH still remained acidic indicating an excess of CODs. Crab shell contains calcium carbonate (Arvanitoyannis and Kassaveti, 2008) and it mainly comprises of inorganic solids (79%, Table 8-1). The pH increased to 7.1-7.5 in experiments performed using crab shell, both in the hydrolysis-fermentation and sulphate reduction stage, most likely due to the dissolution of calcium carbonate. The CODs produced in the experiments did not reach concentrations > 2000 $mg.L^{-1}$ (Figure 8-3), which could be an inhibiting factor according to the literature (Siegert and Banks, 2005). Therefore, the possible inhibition of the anaerobic processes was discarded.

The sulphate reduction profiles with different CBP (Figure 8-4) followed first order kinetics, the specific sulphate removal rates (mg $SO_4^{2-}.VSS^{-1}d^{-1}$) were in the following order: 0.065-0.050 for 2 and 5 mm^2 filter paper > 0.022-0.034 for 2 and 5 mm^3 potato > 0.006-0.009 for 2 and 4 mm Ø crab shells (Table 8-3). The sulphate removal rates are based in the CODs dependency by SRB (Bernardez et al., 2013; Chou et al., 2008; Moosa et al., 2002). The sulphate removal efficiencies using potato and filter paper as CBP (≥ 98%) were in the level of those reported in the literature for other carbon sources, e.g. with increasing CODs bioavailability: cheese whey (94%), wine industry waste (95%), municipal wastewater (>80%) and horse manure (61%) (Castillo et al., 2012; Deng and Lin, 2013; Martins et al., 2009). The rates are related to the nature and the degree of polymerization of the carbohydrates (Jeihanipour et al., 2011; Labatut et al., 2011). Also, it has been recommended that a mixture of fast and slow degradable organic wastes facilitate the sulphate removal efficiency (Kijjanapanich et al., 2012), e.g. the experiments with crab shell showed the lowest sulphate reduction efficiencies and rates in this research, even though the COD as COD_R content in crab shell was 0.53 g COD_R.g VS^{-1}, besides the VS content was 19 % (Table 8-1) which is well within the range reported in the literature (Arvanitoyannis and Kassaveti, 2008). The sulphate reduction efficiency was 32-45% using crab shell even when this CBP can provide sufficient electron donor to carry out the sulphate reduction process. Hence, the semi-crystalline structure of chitin (Pillai *et al.*, 2009) makes it difficult to solubilise and, therefore, it is not easily available for microorganisms. This characteristic makes crab shell a recalcitrant organic waste (O'Keefe *et al.*, 1996). On the other hand, crystallinity (degree of polymerization) is a factor affecting also cellulose: highly crystalline cellulose hydrolyses and ferments at lower rates when compared to amorphous cellulose (Jeihanipour *et al.*, 2011).

Furthermore, in this research, inhibition due to sulphide was discarded. The concentration of sulphide produced in all sulphate reducing experiments (≤ 125 mg $S^{2-}.L^{-1}$) did not inhibit the bacteria. SRB are sulphide tolerant (100-800 mg $S^{2-}.L^{-1}$) in comparison to methane producing archaea that have a lower threshold (McCartney and Oleszkiewicz, 1991).

8.4.3 Implications of CBP for biological sulphate reduction

This work not only aimed to reduce sulphate to sulphide at the fastest rate, but also to find out the process parameters, *e.g.* supply of electron donor to SRB at the lowest rate, which could yield the highest sulphate reduction rate, the optimal utilization of the CODs released by SRB and, in this way, to decrease the cost of the biological sulphate removal process.

We do not recommend the use of high quality potatoes or expensive pure cellulose (based filter paper) as feedstock for the anaerobic sulphate reduction bioprocesses, but rather low quality products as potato peels or waste paper. However, these might contain a considerable fraction of lignocellulosic materials and thus give rise to lower reaction rates. This study gave, nevertheless, an understanding of the reaction kinetics to help to elucidate whether slow release compounds are able to meet the electron donor demand of SRB. Further research should be done with *e.g.* larger particle sizes and/or more complex biopolymers, *e.g.* lignocellulosic biowastes. Also, the use of CBP in sulphate reducing bioreactors, *e.g.* in continuous inverse fluidized bed reactors, can be investigated in order to intensify the sulphate reduction process.

Although this research focused on identifying CBP as SRED for SRB, it also provides an insight into the practicality of coupling sulphate reduction to metal removal in wastewater. Recently, interest has grown for processes which involve bacterial sulphate reduction to sulphide and simultaneous precipitation of metals *in situ* by utilization of the sulphide (Lewis, 2010), even allowing selective metal precipitation (Sampaio *et al.*, 2009). Settlability is a major factor governing the quality of crystals formed during precipitation and, under such conditions, the sulphide concentration in the reactor influences the final product (Villa-Gomez *et al.*, 2012). To avoid supersaturation, electron donors must be delivered slowly to the SRB and thus generate sulphide concentrations that allow the formation of crystalline, well setting metal sulphide precipitates.

8.5 Conclusions

Sulphate reduction using CBP as SRED performed at different specific rates (mg SO_4^{2-}.VSS^{-1}d^{-1}): 0.065-0.050 for 2 and 5 mm^2 filter paper, > 0.022-0.034 for 2 and 5 mm^3 potato cubes and > 0.006-0.009 for 2 and 4 mm Ø crab shells. The differences in the sulphate reduction rates were influenced by the CBP specific hydrolysis-fermentation rates (mg COD$_S$.mg VSS^{-1}d^{-1}): 0.087-0.070 for 2 and 5 mm^3 potato > 0.039-0.047 for 2 and 5 mm^2 filter paper > 0.011-0.028 for 2 and 4 mm Ø crab shells. The hydrolysis-fermentation was the step limiting for using CBP based SRED for sulphate reduction. Therefore, the nature and the degree of polymerization of the CBP affected the sulphate reduction efficiencies. Filter paper offered the best sulphate reduction (> 98%) compared to potato cubes (> 82%) and crab shell (> 32%). The complexity of chitin in crab shell controls the low hydrolysis-fermentation rate, the release of electron donor (0.011-0.028 mg COD$_S$.mg VSS^{-1}d^{-1} for 2 and 4 mm Ø) and, hence, the sulphate reduction rate (0.006-0.009 mg SO_4^{2-}.VSS^{-1}d^{-1} for 2 and 4 mm Ø). The selection of the appropriate SRED needs to be tailored as a function of the sulphate concentration of the inorganic wastewater.

8.6 References

Angelidaki, I., Alves, M., Bolzonella, D., Borzacconi, L., Campos, J.L., Guwy, A.J., Kalyuzhnyi, S., Jenicek, P., van Lier, J.B., 2009. Defining the biomethane potential (BMP) of solid organic wastes and energy crops: a proposed protocol for batch assays. Water Sci. Technol. 59, 927–934. doi:10.2166/wst.2009.040

APHA, 1999. Standard Methods for the Examination of Water and Wastewater, 20th ed. Washington, USA.

Arvanitoyannis, I.S., Kassaveti, A., 2008. Fish industry waste: treatments, environmental impacts, current and potential uses. Int. J. Food Sci. Technol. 43, 726–745. doi:10.1111/j.1365-2621.2006.01513.x

Bai, H., Kang, Y., Quan, H., Han, Y., Sun, J., Feng, Y., 2013. Treatment of acid mine drainage by sulfate reducing bacteria with iron in bench scale runs. Bioresour. Technol. 128, 818–822. doi:10.1016/j.biortech.2012.10.070

Bernardez, L.A., de Andrade Lima, L.R.P., de Jesus, E.B., Ramos, C.L.S., Almeida, P.F., 2013. A kinetic study on bacterial sulfate reduction. Bioprocess Biosyst. Eng. 36, 1861–1869. doi:10.1007/s00449-013-0960-0

Borrego, J., Carro, B., López-González, N., de la Rosa, J., Grande, J.A., Gómez, T., de la Torre, M.L., 2012. Effect of acid mine drainage on dissolved rare earth elements geochemistry along a fluvial–estuarine system: the Tinto-Odiel Estuary (S.W. Spain). Hydrol. Res. 43, 262–274. doi:10.2166/nh.2012.012

Castillo, J., Pérez-López, R., Sarmiento, A.M., Nieto, J.M., 2012. Evaluation of organic substrates to enhance the sulfate-reducing activity in phosphogypsum. Sci. Total Environ. 439, 106–113. doi:10.1016/j.scitotenv.2012.09.035

Chevalier-Billosta, V., Joseleau, J., Cochaux, A., Ruel, K., 2006. Tying together the ultrastructural modifications of wood fibre induced by pulping processes with the mechanical properties of paper. Cellulose 14, 141–152. doi:10.1007/s10570-006-9081-0

Chou, H.-H., Huang, J.-S., Chen, W.-G., Ohara, R., 2008. Competitive reaction kinetics of sulfate-reducing bacteria and methanogenic bacteria in anaerobic filters. Bioresour. Technol. 99, 8061–7. doi:10.1016/j.biortech.2008.03.044

de Lemos Chernicharo, C.A., 2007. Biological Wastewater Treatment Vol.4: Anaerobic Reactors, 1st ed, Biological wastewater treatment in warm climate regions. IWA Publishing, London, UK.

Deng, D., Lin, L.-S., 2013. Two-stage combined treatment of acid mine drainage and municipal wastewater. Water Sci. Technol. 67, 1000–1007. doi:10.2166/wst.2013.653

Houbron, E., González-López, G.I., Cano-Lozano, V., Rustrían, E., 2008. Hydraulic retention time impact of treated recirculated leachate on the hydrolytic kinetic rate of coffee pulp in an acidogenic reactor. Water Sci. Technol. 58, 1415–1421. doi:10.2166/wst.2008.492

Jeihanipour, A., Niklasson, C., Taherzadeh, M.J., 2011. Enhancement of solubilization rate of cellulose in anaerobic digestion and its drawbacks. Process Biochem. 46, 1509–1514. doi:10.1016/j.procbio.2011.04.003

Kijjanapanich, P., Pakdeerattanamint, K., Lens, P.N.L., Annachhatre, A.P., 2012. Organic substrates as electron donors in permeable reactive barriers for removal of heavy metals from acid mine drainage. Environ. Technol. 33, 2635–2644. doi:10.1080/09593330.2012.673013

Labatut, R.A., Angenent, L.T., Scott, N.R., 2011. Biochemical methane potential and biodegradability of complex organic substrates. Bioresour. Technol. 102, 2255–2264. doi:10.1016/j.biortech.2010.10.035

Lakaniemi, A.M., Nevatalo, L.M., Kaksonen, A.H., Puhakka, J.A., 2010. Mine wastewater treatment using *Phalaris arundinacea* plant material hydrolyzate as substrate for sulfate-reducing bioreactor. Bioresour. Technol. 101, 3931–3939. doi:10.1016/j.biortech.2010.01.020

Lenihan, P., Orozco, A., ONeill, E., Ahmad, M.N.M., Rooney, D.W., Mangwandi, C., Walker, G.M., 2011. Kinetic modelling of dilute acid hydrolysis of lignocellulosic biomass, in: Dos Santos Bernardes, M.A. (Ed.), Biofuel production-recent developments and prospects. InTech, pp. 293–308. doi:10.5772/17129

Lewis, A.E., 2010. Review of metal sulphide precipitation. Hydrometallurgy 104, 222–234. doi:10.1016/j.hydromet.2010.06.010

Liamleam, W., Annachhatre, A.P., 2007. Electron donors for biological sulfate reduction. Biotechnol. Adv. 25, 452–463. doi:10.1016/j.biotechadv.2007.05.002

Martins, M., Faleiro, M.L., Barros, R.J., Veríssimo, A.R., Costa, M.C., 2009. Biological sulphate reduction using food industry wastes as carbon sources. Biodegradation 20, 559–567. doi:10.1007/s10532-008-9245-8

McCartney, D.M., Oleszkiewicz, J.A., 1991. Sulfide inhibition of anaerobic degradation of lactate and acetate. Water Res. 25, 203–209. doi:10.1016/0043-1354(91)90030-T

Monterroso, C., Macías, F., 1998. Drainage waters affected by pyrite oxidation in a coal mine in Galicia (NW Spain): Composition and mineral stability. Sci. Total Environ. 216, 121–132. doi:10.1016/S0048-9697(98)00149-1

Moosa, S., Nemati, M., Harrison, S.T.L., 2002. A kinetic study on anaerobic reduction of sulphate, Part I: Effect of sulphate concentration. Chem. Eng. Sci. 57, 2773–2780. doi:10.1016/S0009-2509(02)00152-5

Nieto, J.M., Sarmiento, A.M., Olías, M., Canovas, C.R., Riba, I., Kalman, J., Delvalls, T.A., 2007. Acid mine drainage pollution in the Tinto and Odiel rivers (Iberian Pyrite Belt, SW Spain)

and bioavailability of the transported metals to the Huelva Estuary. Environ. Int. 33, 445–455. doi:10.1016/j.envint.2006.11.010

O'Keefe, D.M., Owens, J., Chynoweth, D., 1996. Anaerobic composting of crab-picking wastes for byproduct recovery. Bioresour. Technol. 58, 265–272. doi:10.1016/S0960-8524(96)00104-6

Park, S.-J., Yoon, T.-I., Bae, J.-H., Seo, H.-J., Park, H.-J., 2001. Biological treatment of wastewater containing dimethyl sulphoxide from the semi-conductor industry. Process Biochem. 36, 579–589. doi:10.1016/S0032-9592(00)00252-1

Pedreschi, F., Travisany, X., Reyes, C., Troncoso, E., Pedreschi, R., 2009. Kinetics of extraction of reducing sugar during blanching of potato slices. J. Food Eng. 91, 443–447. doi:10.1016/j.jfoodeng.2008.09.022

Pillai, C.K.S., Paul, W., Sharma, C.P., 2009. Chitin and chitosan polymers: Chemistry, solubility and fiber formation. Prog. Polym. Sci. 34, 641–678. doi:10.1016/j.progpolymsci.2009.04.001

Pokhrel, D., Viraraghavan, T., 2004. Treatment of pulp and paper mill wastewater - a review. Sci. Total Environ. 333, 37–58. doi:10.1016/j.scitotenv.2004.05.017

Pruden, A., Messner, N., Pereyra, L., Hanson, R.E., Hiibel, S.R., Reardon, K.F., 2007. The effect of inoculum on the performance of sulfate-reducing columns treating heavy metal contaminated water. Water Res. 41, 904–14. doi:10.1016/j.watres.2006.11.025

Rengaraj, S., Joo, C.K., Kim, Y., Yi, J., 2003. Kinetics of removal of chromium from water and electronic process wastewater by ion exchange resins: 1200H, 1500H and IRN97H. J. Hazard. Mater. 102, 257–275. doi:10.1016/S0304-3894(03)00209-7

Robles-González, V., Galíndez-Mayer, J., Rinderknecht-Seijas, N., Poggi-Varaldo, H.M., 2012. Treatment of mezcal vinasses: a review. J. Biotechnol. 157, 524–46. doi:10.1016/j.jbiotec.2011.09.006

Sampaio, R.M.M., Timmers, R.A., Xu, Y., Keesman, K.J., Lens, P.N.L., 2009. Selective precipitation of Cu from Zn in a pS controlled continuously stirred tank reactor. J. Hazard. Mater. 165, 256–265. doi:10.1016/j.jhazmat.2008.09.117

Siegert, I., Banks, C., 2005. The effect of volatile fatty acid additions on the anaerobic digestion of cellulose and glucose in batch reactors. Process Biochem. 40, 3412–3418. doi:10.1016/j.procbio.2005.01.025

Vaccari, D.A., Strom, P.F., Alleman, J.E., 2005. Microbial Transformations, in: Environmental Biology for Engineers and Scientists. John Wiley & Sons, Inc., Hoboken, NJ, USA, pp. 387–441. doi:10.1002/0471741795.ch13

Villa-Gomez, D., Ababneh, H., Papirio, S., Rousseau, D.P.L., Lens, P.N.L., 2011. Effect of sulfide concentration on the location of the metal precipitates in inversed fluidized bed reactors. J. Hazard. Mater. 192, 200–207. doi:10.1016/j.jhazmat.2011.05.002

Villa-Gomez, D.K., Papirio, S., van Hullebusch, E.D., Farges, F., Nikitenko, S., Kramer, H., Lens, P.N.L., 2012. Influence of sulfide concentration and macronutrients on the characteristics of metal precipitates relevant to metal recovery in bioreactors. Bioresour. Technol. 110, 26–34. doi:10.1016/j.biortech.2012.01.041

Wang, A., Ren, N., Wang, X., Lee, D., 2008. Enhanced sulfate reduction with acidogenic sulfate-reducing bacteria. J. Hazard. Mater. 154, 1060–1065. doi:10.1016/j.jhazmat.2007.11.022

Yentekakis, I. V., 2006. Open- and closed-circuit study of an intermediate temperature SOFC directly fueled with simulated biogas mixtures. J. Power Sources 160, 422–425. doi:10.1016/j.jpowsour.2005.12.069

Chapter 9
Lignocellulosic biowastes as carrier material and slow release electron donor for sulphidogenesis of wastewater in an inverse fluidized bed bioreactor

A modified version of this chapter was published as:
Reyes-Alvarado L.C., Camarillo-Gamboa Á., Rustrian E., Rene E.R., Esposito G., Lens P.N.L., Houbron E., (2017). Lignocellulosic biowastes as carrier material and slow release electron donor for sulphidogenesis of wastewater in an inverse fluidized bed bioreactor, Environ. Sci. Pollut. Res. doi:10.1007/s11356-017-9334-5.

Abstract

Industrial wastewaters containing high concentrations of sulphate, such as those generated by mining, metallurgical and mineral processing industries, require electron donor for biological sulfidogenesis. In this study, five types of lignocellulosic biowastes were characterized as potential low cost slow release electron donors for application in a continuously operated sulphidogenic inverse fluidized bed bioreactor (IFBB). Among them natural scourer and cork were selected due to their high composition of volatile solids (VS), *viz.* 89.1 and 96.3%, respectively. Experiments were performed in batch (47 d) and in an IFBB (49 d) using synthetic sulphate-rich wastewater. In batch, the scourer gave higher sulphate reduction rates (67.7 mg SO_4^{2-} L^{-1} d^{-1}) in comparison to cork (12.1 mg SO_4^{2-} L^{-1} d^{-1}), achieving > 82% sulphate reduction efficiencies. In the IFBB packed with the natural scourer, the average sulphate reduction efficiency was 24 ($\pm$ 17)%, while the volumetric sulphate reduction rate was 167 ($\pm$ 117) mg SO_4^{2-} L^{-1} d^{-1}. The long incubation time in the batch experiments (47 d) allowed the higher sulphate reduction efficiencies in comparison to the short hydraulic retention time (24 h) in the IFBB. This suggests the hydrolysis-fermentation was the rate limiting step and the electron donor supply (through hydrolysis of the lignocellulosic biowaste) was limiting the sulphate reduction.

Keywords: Sulphate reducing bacteria (SRB), sulphidogenesis, sulphate reduction, lignocellulosic biowastes, lignocellulosic slow release electron donor (L-SRED), inverse fluidized bed bioreactor (IFBB)

9.1 Introduction

Industrial wastewaters rich in sulphate, such as mining and electronic industry wastewaters and acid mine drainage, are often characterized by a low pH and high redox potential, containing toxic metals and no or low concentrations of organic matter (Table 9-1). When inappropriate treatment is provided, processes related to mining and electronic device production polluted water, soil, air and sediments. In most cases, mining wastewater is not adequately treated and this often leads to uncontrolled sulphide emissions that contribute to odour nuisance and corrosion (Kump *et al.* 2005). There are, to the best of our knowledge, no regulations on the maximal permitted sulphate concentrations in these industrial wastewaters. Some drinking water guidelines nevertheless refer to maximal allowable sulphate concentrations, *e.g.* the Mexican regulation for drinking water (NOM-127-SSA1, 1994) allows a maximum sulphate concentration of 400 mg L^{-1}, the US Environmental Protection Agency (EPA, 1994) recommends a maximum sulphate concentration of 250 mg L^{-1} and the World Health Organization (WHO, 1994) suggests the maximum sulphate concentration of drinking water should not exceed 500 mg L^{-1}.

Biological sulphate removal is usually carried out by sulphate reducing bacteria (SRB), which reduce sulphate to sulphide under anaerobic conditions (Grein *et al.* 2013). The latter precipitates with the heavy metals present in the mining wastewater and residual sulphide, if any, is partially oxidized to elemental sulphur (S^0) that can be removed from the industrial wastewater (Lens *et al.* 2002). The sulphate reducing capability of SRB is utilized in the first step of biological sulphate removal (sulphidogenesis: SO_4^{2-} reduction to S^{2-}) from wastewater, and is applied in many different bioreactor configurations, from extensive systems as lagoons and wetlands to high rate anaerobic bioreactors. A detailed description of these different bioreactors performing sulphate reduction can be found elsewhere in the literature (Kaksonen and Puhakka, 2007).

Table 9-1. Physico-chemical characteristics of acid mine drainage (AMD) from different regions

Origin (country)	pH	SO_4^{2-} (mg L^{-1})	EC** (mS cm^{-1})	E_h*** (mV)	Metals	COD (mg L^{-1})	References
Spain (As Puentes lignite mine dump Galicia, Spain was)	2.2-8.0	424-7404	0.82-6.51	143-754	F, Ca, Mg, Na, K, Si, Al, Fe, Mn, Ni, Co, Zn, Cu, Pb and Cd.		(Monterroso and Macías 1998)
Spain (The Tinto and Odiel Rivers-estuarine systems)	< 3		17.31		Sc, Y, U, and La, Ce, Pr, Nd, Sm, Eu, Gd, Tb, Dy, Ho, Er, Yb, Lu, REE*		(Borrego *et al.* 2012)
Spain (The Tinto River)	2.3	8500			Fe, Cu, Zn, Al, Mn, Ni, Cd and Cr		(Jiménez-Rodríguez *et al.* 2009)
Portugal (S. Domingos, Alentejo)	~ 2.0	3100			K, Ca, Ti, V, Cr, Mn, Fe, Ni, Cu, Zn, As, Se, Rb, Sr, Cd, Sn, Sb, Ba and Pb		(Martins *et al.* 2009)
USA (Dunkard Creek downstream of Taylortown, Pennsylvania)	4.2 (± 0.9)	1846 (± 594)	21980 (± 4870)		Fe, Ca, Mg, Mn, Al and Na	41 (± 49)	(Deng and Lin 2013)
India (Mines: Baragolai, Ledo, Tikak, Tirap and Tipong)	2.3-7.6	176–3615	0.785-6.76		Na, K, Ca, Mg, Cr, Ni, Zn, Mn, Fe, Al, Cd, Pb, Cu and Co		(Equeenuddin *et al.* 2010)
China (copper mine)	2.8	20800			Cu, Fe and Mn	< 100	(Bai *et al.* 2013)
France (Chessy-Les-Mines)	2.6	5800			Fe, Zn, Cu, Al, Mn, Co, Ni and Pb		(Foucher *et al.* 2001)
Zimbabwe (Iron-Duke Mine in Glendale)	2.1-2.7	16300-19000	11.4-14.1		As, Ni and Fe		(Mapanda *et al.* 2007)

Note: *Rare Earth Elements (lanthanide series), **EC = electrical conductivity, ***E_h = redox potential

The use of vast extensions of land represents an economical and unsustainable disadvantage for sulphate removal from industrial wastewater in natural treatment systems. Therefore, anaerobic bioreactors with small size are preferred over wetlands. Furthermore, in anaerobic bioreactors, it is possible to concentrate a large population of either autotrophic (Sipma *et al.* 2004; Parshina *et al.* 2005a; Parshina *et al.* 2005b; Sipma *et al.* 2007) or heterotrophic (Lopes *et al.* 2007; Sarti and Zaiat 2011; Cao *et al.* 2012) SRB. Processes using autotrophic bacteria can lower the cost of bioremediation (Matassa *et al.* 2014) when simple substrates such as carbon dioxide or hydrogen are used. When using heterotrophic SRB, the use of a waste as substrate can lower the overall cost of biological sulphate reduction. Several complex substrates have already been tested for anaerobic sulphate removal, such as cheese whey and wastes from the wine industry (Martins *et al.* 2009b), leachate and residues resulting from the chemical acid hydrolysis of the plant *Phalaris arundinacea* (Lakaniemi *et al.* 2010), horse manure, vegetable skins, legume compost (Castillo *et al.* 2012), rice husk filtrate (Chockalingam *et al.* 2005), landfill leachate (Thabet *et al.* 2009), high molecular weight lignin (Ko *et al.* 2009) and molasses (Teclu *et al.* 2009).

The inverse fluidized bed bioreactor (IFBB) has been widely studied for methanogenic processes, such as for COD removal from wine distillery wastewater (Garcia-Calderon *et al.* 1998). Recently, there has been effort to optimize the simultaneous removal of carbon and nitrogen in this type of bioreactors (Alvarado-Lassman *et al.* 2006). Sulphate removal from wastewater using an IFBB is a promising technology because it can achieve high sulphate reduction efficiencies (Celis-García *et al.* 2007) and two important processes can be combined in one reactor: i) sulphidogenesis and ii) subsequent metal sulphide precipitation (Villa-Gomez *et al.* 2011). Although IFBB have been used for sulphate removal from industrial wastewater (Villa-Gomez *et al.* 2011; Kijjanapanich *et al.* 2014; Janyasuthiwong *et al.* 2016), this bioreactor type deserves more attention for further optimization and intensification of the biological sulphate reduction process.

Supply of an external, pure and expensive carbon source and electron donor to industrial wastewaters that have low chemical oxygen demand (COD) concentrations to support the growth and activity of heterotrophic SRB can make the anaerobic treatment of these wastewater types economically unattractive or unsustainable for use in developing countries. Differently, waste materials containing a high organic matter content can be used as slow release electron donors (SRED) and could thus allow cost-effective biological sulphate reduction. Good candidates are waste residues from agriculture and forestry, as photosynthesis produces abundant amounts of biomass: lignocellulosic biomass is produced at the rate of ~ 170×10^9 t of biomass year^{-1} (Kamm and Kamm 2004; Corma *et al.* 2007). It comprises of major constituents like cellulose, hemicellulose and lignin, and minor constituents like terpenes, resins, colours and tannins. The final composition differs between the sources depending on the climate and geographical location (Malherbe and Cloete 2002; Pérez *et al.* 2002).

From a process intensification point of view, research on the application of high rate feeding conditions, the use of organic solid wastes (*e.g.* lignocellulose) as SRED, SRED as carrier material and the optimization of operating conditions in an IFBB to maintain high sulphate reduction efficiencies by biomass should be undertaken to develop a cost-effective biological sulphate reduction process in an IFBB. Reyes-Alvarado *et al.* (2017) investigated the use of carbohydrate based polymeric materials as SRED (starch, cellulose and chitin) as potential electron donors for a sulphidogenic process in batch bioreactors. In this study, lignocellulosic biowaste was investigated for the capability to release COD to a synthetic wastewater in the absence (natural release) and in the presence (hydrolysis-fermentation) of inoculum. Moreover, the sulphate reduction process using the slowly released COD from L-SRED was investigated in batch and a continuously operated IFBB bioreactors.

9.2   Material and methods

9.2.1   Inoculum

The inoculum was obtained from a pilot scale upflow anaerobic sludge blanket (UASB) reactor treating cheese whey, located at the Laboratory of Environmental Biotechnology, Faculty of Chemistry, Universidad Veracruzana (Orizaba, Ver., Mexico). The seeding liquid contained 22.4 ($\pm$ 0.30) g L^{-1} total suspended solids (TSS) and 16.4 ($\pm$ 0.28) g L^{-1} volatile suspended solids (VSS).

The biomass used for the batch reactors was prepared as follows: first it was centrifuged (5000$\times$g for 10 min) and the supernatant was removed. Afterwards, the biomass was re-suspended in synthetic wastewater (free of any electron donor or acceptor). Finally, it was again centrifuged (5000$\times$g for 10 min) in order to remove the supernatant. This procedure was done for all batch experiments to remove any soluble COD (COD$_S$) coming from the pilot scale UASB reactor. The concentrated biomass was again re-suspended in synthetic wastewater and added to the respective batch incubations. This procedure was necessary to keep the initial inoculum concentration (2 g VSS L^{-1}) constant in all batch experiments. The two continuous IFBB were inoculated with 10% (*v/v*) of sludge that was directly added after collection from the pilot scale UASB reactor.

9.2.2   Synthetic wastewater composition

The composition of the synthetic wastewater used in this study consisted of (in mg L^{-1}): NH$_4$Cl (300), MgCl$_2$·6H$_2$O (120), KH$_2$PO$_4$ (200), KCl (250), CaCl$_2$·2H$_2$O (15), yeast extract (20) and

0.5 mL of a mixture of micronutrients. The micronutrients had the following composition (in mg L^{-1}): FeCl$_2$·4H$_2$O (1500), EDTA (500), H$_3$BO$_3$ (62), NaMoO$_4$·2H$_2$O (36) and HCl 36 % (1 mL) (Villa-Gomez *et al.* 2011). Ethanol (only used in sulphate reducing activity tests of the inoculum) and sodium sulphate were used as, respectively, the electron donor and acceptor when needed. All reagents used in this study were of analytical grade.

9.2.3 Lignocellulose as SRED

Lignocellulosic wastes such as banana peels, natural scourer (*Curcubitaceae* family and *Luffa* genus (Tanobe *et al.*, 2005)), sugarcane leaves, coffee pulp and cork were characterized and screened as potential lignocellulosic SRED (L-SRED). The scourer and cork were selected for sulphate reduction experiments. The natural scourer was dried at 100 °C for 24 h and cut into thin strips of ~ 2 cm length and 0.3 g was added to each batch bioreactor. In the IFBB, a single piece of the same natural scourer (13 cm length, ~ 2.6 cm of diameter, similar to the IFBB internal diameter) with a total mass of 2.4 g dry weight was used. The cork was crushed and the particle size ranged between 0.8-1 mm in diameter. Afterwards it was also dried at 100 °C for 24 h, 0.3 g and 2.4 g were used in the batch and IFBB, respectively.

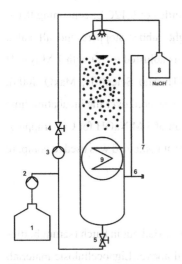

Figure 9-1. Schematic of the inverse fluidized bed bioreactor
Main components: 1) Influent tank, 2) Peristaltic pump, 3) Recirculation pump, 4) Recirculation control valve, 5) Safety valve, 6) Sampling area, 7) Effluent connected to the sewage pipe, 8) Gas trap and 9) Heating coil to control the temperature using a water bath. The small spheres inside the column represent either the scourer or the cork which were used as the carrier material and the slow release electron donor, respectively.

9.2.4   Anaerobic IFBB set up

Two identical IFBB were built using transparent polyvinyl chloride (PVC) pipes (internal diameter = 2.8 cm and length = 27 cm). The effective working volume of the two IFBB was 0.153 L, corresponding to 25 cm of column height (Figure 9-1). The influent was supplied to the system with the help of a peristaltic pump (Masterflex L/S), installed before the recirculation pump (Masterflex L/S). The wastewater inside the IFBB was recirculated downwards the column. The effluent outlet was placed at a distance of 5 cm from the bottom of each IFBB. The temperature (30 °C) was controlled with the help of a heating coil, containing warm water generated from a water bath (Cole Parmer, Polystat 12112-00), that surrounded the reactors. The coils were protected externally with polyethylene foam in order to prevent heat loss.

9.2.5   Sulphate reducing and methanogenic activity tests with anaerobic sludge

The methanogenic activity of the inoculum using ethanol (1 g $COD_{Ethanol}$ L^{-1}) as the substrate was determined as described by Angelidaki *et al.* (2009). The sulphate reduction activity of the inoculum was determined by using ethanol (1.3 g $COD_{Ethanol}$ L^{-1}) and sulphate (0.7 g SO_4^{2-} L^{-1}) (initial COD:sulphate ratio of 1.8), adopting the procedure described by Villa-Gomez *et al.* (2011). The batch experiments were performed in serum bottles of 0.120 L, containing 0.1 L synthetic wastewater. The bottles were fitted with airtight rubber stoppers and all batch experiments were performed in triplicates. The initial pH was adjusted to 7.0 with 1 M NaOH and all batch bottles were placed on an orbital shaker (Thermo Scientific, MaxQ 4000), maintained at 160 rpm and 30 °C. Methane production was measured over the incubation time using an inverted measuring cylinder that contained a solution of 3 M NaOH for CO_2 trapping. Sulphate concentration profiles were also monitored and the data obtained was used to compute the respective activities.

9.2.6   L-SRED and sulphate reduction experiments

The COD_S release and sulphate reduction experiments were carried out in batch (serum bottles with 120 mL capacity) using the same conditions described above. Lignocellulosic material, either the natural scourer or cork (0.3 g dry weight), were added as L-SRED. Apart from the L-SRED, no external soluble electron donor was added. The L-SRED were investigated as follows: Test 1) COD_S release in synthetic wastewater under anaerobic conditions without inoculum and under non-sterile conditions, Test 2) COD_S release with inoculum (2 g VSS L^{-1}, hydrolysis-fermentation) and Test 3) COD_S release and sulphate reduction (700 mg SO_4^{2-} L^{-1} in the synthetic wastewater) with inoculum (2 g VSS L^{-1}).

9.2.7 L-SRED experiments in IFBB

Two identical IFBB were loaded with the scourer and the cork (as indicated above) as L-SRED, inoculated and filled with the synthetic wastewater containing sulphate (700 mg SO_4^{2-} L^{-1}). The inoculum was added at 10 % of the active volume of the IFBB, $e.g.$ 0.015 L. The IFBB were operated in continuous mode from the first day at 30 °C and a constant hydraulic retention time (HRT) of 1 d. Sulphate was added continuously at a concentration of 700 mg L^{-1} in the influent. No external soluble electron donor was added apart from the L-SRED. CODs, VSS, total dissolved sulphide (TDS), sulphate and pH were monitored over the operation time every second day.

9.2.8 Qualitative assessment of SRB growth

Solid samples from the carrier material (L-SRED) used for sulphate reduction in the IFBB and samples from the liquid effluent of the IFBB were investigated for the presence of SRB according to the protocol described elsewhere (Iverson 1966). In this test, the biomass from the L-SRED (attached growth) and the liquid effluent from the IFBB (suspended growth) were grown on an agar plate containing tripticase soy agar (40 g L^{-1}), additional agar (5 g L^{-1}), sodium lactate at 60% (0.4% v/v), hydrated magnesium sulphate (0.5 g L^{-1}) and ferrous ammonium sulphate (0.5 g L^{-1}). The pH of the agar medium was adjusted to 7.2-7.4 with 1 M NaOH solution and incubated at 30 °C. Samples from the reactor were taken with sterile forceps and rinsed with sterile deionized water to remove the non-attached biomass. This pre-washing step was done to differentiate the suspended biomass from the attached biomass growing on the L-SRED.

9.2.9 Calculations

The volumetric removal rates in batch (V_r) and specific removal rates (S_r) in batch, corresponding to COD or sulphate, the volumetric removal rate in the IFBB ($V_{r\text{-}IFBB}$) and the sulphate removal efficiency (SRE), were calculated using the following equations:

$$V_r = \frac{(y_0 - y_1)}{(t_1 - t_0)} = \text{mg L}^{-1}\text{d}^{-1} \qquad \text{Eq. 9-1}$$

$$S_r = \frac{V_r}{[VSS]} = \text{mg mg VSS}^{-1}\text{d}^{-1} \qquad \text{Eq. 9-2}$$

$$V_{r\text{-}IFBB} = \frac{Q_{in}(y_0 - y_1)}{(V_{reactor})} = \text{mg L}^{-1}\text{d}^{-1} \qquad \text{Eq. 9-3}$$

$$SRE = \left(\frac{(SO_{4\,initial}^{2-} - SO_{4\,final}^{2-})}{SO_{4\,initial}^{2-}}\right) \times 100 = \% \qquad \text{Eq. 9-4}$$

where, y_0 is the concentration (mg L^{-1}) of a parameter (COD or sulphate) at the beginning of the experiment (t_0, d) and y_1 is the concentration (mg L^{-1}) of the same parameter after the time of reaction (t_1, d). The *VSS* is the concentration (mg L^{-1}) of volatile suspended solids. The volume of the IFBB ($V_{reactor}$) and the influent flow rate (Q_{in}) were considered for the V_{r-IFBB} calculation of the IFBB experiment.

9.2.10 Analytical procedures

The COD, total solids (TS), volatile solids (VS), VSS, TSS, TDS (by the methylene blue reaction) and sulphate concentrations were measured according to the procedure outlined in Standard Methods (APHA 1999). The pH was measured with a pH electrode (Prosense, Oosterhout, The Netherlands). The volatile fatty acids (VFA) such as acetate (C_2), propionate (C_3), butyrate (C_4), isobutyrate (C_{4i}), valerate (C_5) and isovalerate (C_{5i}) were analyzed by gas chromatography coupled to a flame ionization detector (GC-FID, Agilent 6820) and fitted with a DB-FFAP column (0.53 Ø, 30 m, 0.25 µm).

The lignocellulosic biowastes were characterized according to the methods outlined in the Technical Association of the Pulp and Paper Industry standards (TAPPI, 2002). These analyses included cellulose and hemicellulose (TAPPI T203 om-93), lignin (TAPPI T222 om-88) extractable organics with hot water as well as organics with acetone (TAPPI T 264 om-88).

9.3 Results

9.3.1 Characterization of lignocelulosic materials

Five different lignocellulosic materials, *i.e.* banana peels, coffee pulp, sugarcane leaves, cork and scourer, were physico-chemically characterized as shown in Table 9-2. Banana peels, coffee pulp and sugarcane leaves are residues of the food industry. Coffee pulp is often used as an energy source for the roasting step in the wet coffee process, but in most cases it is discarded to the environment without any treatment. However, the discarded coffee pulp still contains caffeine and tannins, resulting in waste disposal problems. Sugarcane residues are usually burned *on site*, after harvesting. Cork can be a waste after opening and utilizing wine bottles. The natural scourer is cultivated and harvested in most tropical areas.

Table 9-2. Characterization of lignocellulosic materials used in this study as L-SRED for sulphate removal

Parameter	Banana peel	Coffee pulp	Sugarcane leaves	Scourer	Cork
TS (%)	12.4	15.8	33.5	92.8	96.9
VS (%)	11.2	14.9	31.0	89.1	96.3
Moisture (%)	87.6	84.2	66.5	7.2	3.1
Ashes (%)	1.2	1.0	2.4	3.7	0.5
Acetone extractable organics (%)	2.8	1.9	2.8	6.9	15.8
Water extractable organics (%)	3.3	5.7	3.6	18.9	1.3
Lignin (%)	2.1	3.0	6.1	10.2	43.8
α-Cellulose (%)	2.8	3.6	18.9	47.5	14.3
β and γ-Cellulose (%)	1.3	1.1	2.2	5.9	15.8

Evidently, the moisture composition is very high in sugarcane leaves (66.5%), banana peels (87.6%) and coffee pulp (84.2%). On the other hand, the natural scourer and cork showed the lowest moisture and the highest TS content, *viz.* 92.8 and 96.9%, respectively (Table 9-2). A high concentration of extractable organics with acetone (15.8%) was observed in the cork. On the other hand, the highest composition of hot water extractable organics (18.9%) was observed in the scourer (Table 9-2).

The cork had the highest lignin content (43.8%), followed by banana peel (10.2%), while the rest of the screened lignocellulosic materials had a lignin content ≤ 6.1 %. The fraction of β and γ-cellulose in cork and the natural scourer was 15.8% and 5.9%, respectively, while that of the other materials was $\leq 2.2\%$. The fraction of α-cellulose in the natural scourer and sugarcane leaves was 47.5% and 18.9%, followed by 14.3% in cork. Hence, the scourer and cork had the highest VS content, thus were selected for the subsequent experiments.

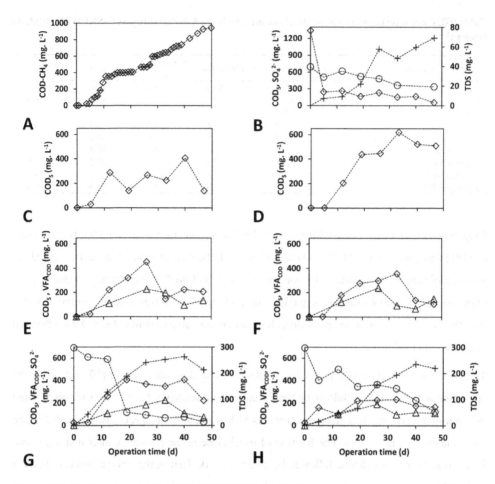

Figure 9-2. Profile of experiments using L-SRED in batch

CODs ($\diamond$), sulphate ($\circ$), sulphide (+, TDS) and total VFA-CODs ($\triangle$). Methanogenic (A) and sulphate reduction (B) activity test. The scourer (C) and cork (D) as L-SRED in non-sterile synthetic wastewater without inoculum and the absence of sulphate addition. Hydrolysis-fermentation of the scourer (E) and cork (F). Sulphate reduction process using scourer (G) and cork (H) as L-SRED.

9.3.2 Methanogenic and sulphate reducing activity of the anaerobic sludge

The tests confirmed SRB, hydrolytic-fermentative bacteria (HFB) and methanogenic archaea (MA) were present in the anaerobic sludge used as inoculum. The inoculum sludge had a specific methanogenic activity of 0.0095 mg COD-CH$_4$ mg VSS^{-1} L^{-1} (Figure 9-2A) and the COD-CH$_4$ formed during the test (927 mg COD-CH$_4$ L^{-1}) was almost equal to the theoretical CODs added as substrate (1000 mg COD$_{Ethanol}$ L^{-1}). The volumetric sulphate-reducing activity of the inoculum (Figure 9-2B) was 7.5 mg SO$_4^{2-}$ L^{-1} d^{-1}, while the specific sulphate reducing activity was 0.0038 mg SO$_4^{2-}$ mg VSS^{-1} d^{-1}. The sulphate removal efficiency was 51% at a COD:sulphate ratio of 1.8 after 47 d of batch incubations.

9.3.3 CODs released from L-SRED in the absence of inoculum (natural release)

The amount of CODs released by the SRED to the synthetic wastewater was measured in the absence of anaerobic sludge and under non-sterile conditions (Figure 9-2C-D). CODs diffused out of both lignocellulosic materials (scourer and cork) into the synthetic wastewater. In the case of the scourer, the CODs concentration ranged from > 140 to < 410 mg COD L^{-1} after 10 days of incubation. When cork was used, the CODs ranged from > 200 and < 630 mg COD L^{-1} after 15 days of incubation. The CODs consisted of extractable organics with acetone and hot water for both L-SRED (Table 9-2).

9.3.4 CODs released from the L-SRED in the presence of inoculum (hydrolysis-fermentation step)

The CODs released from the SRED was quantified in the presence of the inoculum. The CODs observed was ≤ 450 mg COD L^{-1} during the hydrolysis-fermentation step of the scourer and the cork (Figure 9-2E-F). Furthermore, there was VFA production at a concentration ≤ 230 mg COD L^{-1}. In the batch reactors with the natural scourer and the cork, the CODs and VFA profiles followed nearly a similar trend: both increased after 5 d of incubation. This indicates that only the naturally released CODs by the SRED was fermented and the HFB could not further hydrolyse the lignocellulosic material. The initial pH (7.0) decreased to 4.9 ($\pm$ 0.2) and 5.9 ($\pm$ 0.3) for, respectively, the scourer and cork (Table 9-3).

Table 9-3. Kinetic evaluation of L-SRED materials in batch incubations

Experiment	Parameter	Units	L-SRED	
			Scourer	Cork
Natural release using L-SRED	CODs	Vr (mg CODs L^{-1} d^{-1})	36.9	21.2
		Sr (mg CODs mg L-SRED^{-1} d^{-1})	0.12	0.07
Hydrolysis-fermentation using L-SRED	CODs	Vr (mg CODs L^{-1} d^{-1})	20.5	12.7
		Sr (mg CODs mg VSS^{-1} d^{-1})	0.01	0.006
	Total VFA	Vr (mg CODs L^{-1} d^{-1})	8.7	9.2
		Sr (mg CODs mg VSS^{-1} d^{-1})	0.004	0.005
		pH$_{final}$	4.9 ($\pm$ 0.2)	5.9 ($\pm$ 0.3)
Sulphate reduction using L-SRED	CODs	Vr (mg CODs L^{-1} d^{-1})	27.1	10.3
		Sr (mg CODs mg VSS^{-1} d^{-1})	0.014	0.005
	Total VFA	Vr (mg CODs L^{-1} d^{-1})	6.8	7.4
		Sr (mg CODs mg VSS^{-1} d^{-1})	0.003	0.004
	Sulphate	Vr (mg SO_4^{2-} L^{-1} d^{-1})	67.7	12.1
		Sr (mg SO_4^{2-} mg VSS^{-1} d^{-1})	0.034	0.0061
		pH$_{final}$	6.1 ($\pm$ 0.2)	6.8 ($\pm$ 0.2)
		Sulphate removal efficiency (%)	95	82

9.3.5 Sulphate reduction during the release of CODs from the L-SRED in the presence of inoculum (sulphate reduction using L-SRED)

The sulphate removal, at an initial concentration of 700 mg L^{-1}, was studied in batch using the two selected L-SRED (the scourer and cork) and the anaerobic sludge as inoculum (Figure 9-2G-H). Sulphate reduction was possible using both the scourer and cork without the addition of any soluble electron donor. The sulphate reduction efficiency was 95 and 82% using the scourer and cork, respectively. When the scourer was used, three different phases of volumetric and specific sulphate reduction rates were observed: the first one from 0-12 d, the second from 12-19 d and the third from 19-47 d. Among these three rates, the highest rates were observed in the second phase: a V_r of 67.7 mg SO_4^{2-} L^{-1} d^{-1} and a S_r of 0.034 mg SO_4^{2-} mg VSS^{-1} d^{-1} (Table 9-3). Using cork as the L-SRED gave only one rate: a V_r of 12.1 mg SO_4^{2-} L^{-1} d^{-1} and a S_r of 0.0061 mg SO_4^{2-} mg VSS^{-1} d^{-1}.

9.3.6 Production of VFA in batch experiments

In the hydrolysis-fermentation and sulphate reduction experiments using L-SRED, VFA (C_2 to C_5) production was observed. The fraction of each VFA (the COD of the respective VFA was divided by the total amount of VFA at the respective incubation time) produced during the sulphate reduction experiment (Figure 9-2G-H), using the natural scourer and cork as L-SRED, was compared to the fraction of each VFA produced during the hydrolysis-fermentation experiment (Figure 9-2E-F) by linear regression. The slopes of the linear regression analysis were plotted in Figure 9-3; these values indicate accumulation or removal of a certain VFA. The Pearson correlation values (r^2) describe the strong or weak relation between the VFA produced in the presence or absence of sulphate. For instance, in Figure 9-3, a value of 1 indicates that a certain VFA was produced and/or consumed in a similar way during the sulphate reduction or hydrolysis-fermentation step. A value > 1 indicates the accumulation of a certain VFA produced during the sulphate reduction step and/or removed in the hydrolysis-fermentation step, while a value < 1 indicates the inverse.

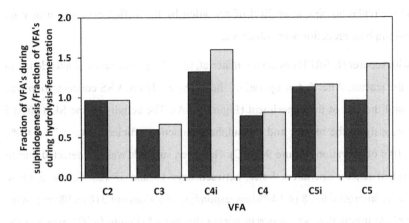

Figure 9-3. Comparison of VFA production in two anaerobic processes in batch (hydrolysis-fermentation and sulphate reduction)
The ratio of VFA produced during the sulphate reduction on the VFA produced during the hydrolysis-fermentation, respectively, using scourer (■) and cork (□) as L-SRED.

Acetate (C_2) was slightly better consumed during the sulphate reduction process in comparison to the hydrolysis-fermentation ($r^2 = 0.97$) using both the scourer and cork as L-SRED (Figure 9-3). Propionate (C_3) concentrations were 40% ($r^2 = 0.98$) and 33% ($r^2 = 0.86$) higher during the use of, respectively, the natural scourer and cork in the hydrolysis-fermentation compared to the sulphate reduction process in batch experiments. The butyrate (C_4) removal was 23% ($r^2 = 0.53$) and 19% ($r^2 = 0.87$) higher during the sulphate reduction process using the scourer and cork, respectively, when compared to the hydrolysis-fermentation of both L-SRED. Iso-butyrate (C_{4i}) accumulated during the sulphate reduction process, 32% ($r^2 = 0.96$) and 60% ($r^2 = 0.83$) for the scourer and cork, respectively, in comparison to the hydrolysis-fermentation process. Likewise, the iso-valerate (C_{5i}) accumulation was 13% ($r^2 = 0.99$) and 16% ($r^2 = 0.93$) more during the sulphate reduction experiments, for the scourer and cork, respectively. Valerate (C_5) was only slightly removed (4%) during the sulphate reduction experiments when the scourer was used as the L-SRED ($r^2 = 0.62$), whereas it accumulated 42% ($r^2 = 0.91$) when cork was used as L-SRED for sulphate reduction (Figure 9-3).

9.3.7 Lignocellulose as carrier material and SRED in an IFBB

9.3.7.1 IFBB performance

Two IFBB were tested for sulphate reduction, *viz.* one using the natural scourer and the second one using cork as the L-SRED. The results with cork in the IFBB were not shown due to operational problems encountered in this study, such as the wash out of the carrier material, clogging of the recirculation pipe after 30 d of operation by the settled cork and nearly no acidification and sulphate reduction were observed.

For the IFBB with scourer (L-SRED) as carrier material, the VSS profile showed a tendency to wash out from the reactor. After 5 d of operation, the average effluent VSS concentration was < 100 mg L^{-1} until the end of the experiment (Figure 9-4A). The activity of the SRB started almost instantaneously in the reactor and the sulphate reduction efficiency was 17 ($\pm$ 10)% during the first 10 d of operation (Figure 9-4B-C). However, sulphide was not detected during this time interval of reactor operation. Later, between days 10 and 33 (23 d), the sulphate reduction efficiency improved to 38 ($\pm$ 14)%, corresponding to a $V_{r\text{-}IFBB}$ of 318 ($\pm$ 98) mg SO$_4^{2-}$ L^{-1} d^{-1}. The sulphide production was apparent during this period (Figure 9-4D), wherein the concentrations reached a maximum of 76 mg L^{-1} on day 27. The average $V_{r\text{-}IFBB}$ observed was 167 ($\pm$ 117) mg SO$_4^{2-}$ L^{-1} d^{-1} and the average sulphate reduction efficiency was 24 ($\pm$ 17)% along the 49 d of operation (Table 9-4).

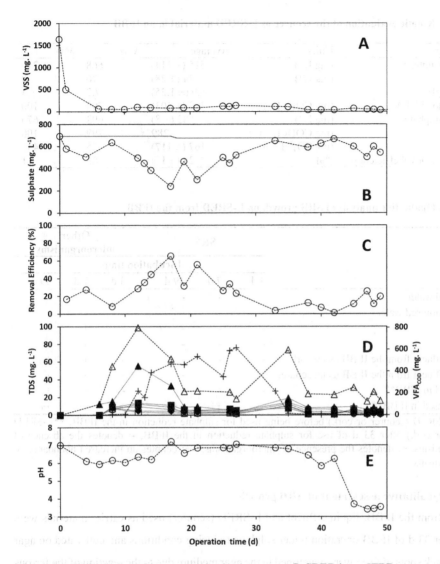

Figure 9-4. Biological sulphate removal using scourer as L-SRED in an IFBB
A) Volatile suspended solids, VSS (○). B) Influent (—) and effluent (○) sulphate. C) Calculated sulphate removal efficiency (○). D) Sulphide production (+), total VFA (△), acetate (▲), propionate (●), iso-butyrate (□), butyrate (■), iso-valerate (○) and valerate (◆). E) Inffluent (—) and effluent (○) pH

Before the sulphidogenesis step (the first 10 d of operation), the total VFA concentration reached a maximum of $\sim$ 800 mg CODs L^{-1} in the IFBB (Fig. 4D). This value was larger than that observed during the periods with sulphide production (261 $\pm$ 143 mg CODs L^{-1}) between days 10 and 33 (Figure 9-4D). After 33 d of operation, the sulphate reduction efficiency fell back to 11 ($\pm$ 8)%, while the sulphide concentrations dropped to almost zero and the effluent total VFA concentration was 241 ($\pm$ 162) mg CODs L^{-1} until the end of the experiment (49 d).

207

Table 9-4. Kinetic evaluation of the scourer as L-SRED material in an IFBB

Parameter	Units	Average	Max	Min
Effluent sulphate	(mg L^{-1})	515 ($\pm$ 114)	668	240
TDS	(mg L^{-1})	24 ($\pm$ 28)	76	0
Effluent pH		5.9 ($\pm$ 1.25)	7.2	3.4
Effluent total VFA	(mg COD$_S$ L^{-1})	289 ($\pm$ 203)	799	100
Influent sulphate	(mg L^{-1})	682 ($\pm$ 8)	693	676
$V_{r\text{-}TVFA}$	(mg COD$_S$ L^{-1} d^{-1})	289	799	100
$V_{r\text{-}IFBB}$	(mg SO$_4^{2-}$ L^{-1} d^{-1})	167 ($\pm$ 117)	25	436
Sulphate removal efficiency	(%)	24 ($\pm$ 17)	65	1

Table 9-5. Qualitative analysis of SRB growth on L-SRED from the IFBB

Sample	SRB			Other microorganisms		
	Incubation time					
	3 d	7 d	49 d	3 d	7 d	49 d
Deionised water	-	-	-	-	+	+
Sterile deionised water	-	-	-	-	-	-
Cork*	-	-	-	-	+	+
Scourer*	-	-	-	-	+	+
Liquid effluent from the IFBB using cork	-	-	-	+	+	+
Liquid effluent from the IFBB using scourer	-	-	-	+	+	+
Cork used in IFBB**	-	-	-	-	+	+
Scourer used in IFBB**	+	++	++	+	++	++

Note: *L-SRED (scourer or cork) before being used for sulphate reduction in the IFBB, **L-SRED (scourer or cork) after 33 d of use for sulphate reduction in the IFBB, -: denotes the absence of microorganisms, +: denotes the presence of microorganisms, ++: denotes an increased abundance of microorganisms

9.3.7.2 Qualitative assessment of SRB growth

Samples from the IFBB, liquid effluent and L-SRED (scourer) used as carrier materials were taken after 33 d of IFBB operation under sulphate reducing conditions and cultivated on agar plates. Black spots of FeS$_2$ were developed in the agar medium due to the reaction of the ferrous ion (present in the agar medium) with the sulphide produced by sulphate reduction. The results of this qualitative microbial test (Table 9-5) demonstrate that SRB were indeed present on the surface of the scourer, as black spots (FeS$_2$) developed on the agar plate within 3 d of incubation. Additionally, some other whitish bacteria also developed on the surface of the agar plates.

9.4 Discussion

9.4.1 Biological sulphate reduction using L-SRED in batch incubations

This study showed that biological sulphate removal by means of dissimilatory reduction is possible using lignocellulosic materials as SRED, *viz.* scourer and cork, in the batch experiments and with scourer in a continuous IFBB. The different sulphate reduction rates obtained in the L-SRED batch experiments were presumably affected by the physico-chemical characteristics of the scourer and the cork (Table 9-2). For instance, the scourer had a higher carbohydrate content, which is easier to degrade when compared to lignin (Ko *et al.* 2009). The three phases of sulphate reduction rates observed using scourer as L-SRED (Figure 9-2G) can be explained as follows: (i) the first phase (0-12 d) was associated with the consumption of the fermentation products of the water extractable organics, (ii) in the second phase (12-19 d), the HFB hydrolyzed and fermented the β and γ - cellulose to make new CODs available, and (iii) during the third phase (19-47 d), the SRB depleted the remaining sulphate with the remainder of the fermented byproducts (CODs). In contrast, the cork showed the highest composition of extractable organics with acetone and lignin; and had a lower carbohydrate content (Table 9-2).

CODs was limited in the synthetic wastewater and the sulphate reduction efficiencies (95% using scourer and 82% using cork) were thus achieved using the electron donors provided by the HFB. HFB are known to produce and excrete exoenzymes such as proteases, lipases and glycosidases (Whiteley and Lee 2006). For instance, the glucosidases are responsible for breaking down long chain carbohydrates, such as starch, chitin, hemicelluloses and celluloses. The permeases transport the hydrolysate by-products (monomers) into the cell, which are subsequently fermented by different catabolic pathways to low molecular weight compounds (CODs) such as VFA and alcohols (Whiteley and Lee 2006).

The CODs concentration released by the L-SRED ($\sim$ 140-620 mg COD L^{-1}, regardless the scourer or cork) to the synthetic wastewater in the absence of inoculum (natural release, Figure 9-2C-D) and the CODs concentration released during the hydrolysis-fermentation process ($\leq$ 450 mg COD L^{-1}, Figure 9-2E-F) did not represent the CODs theoretically required to completely remove the sulphate (700 mg L^{-1}). Accordingly, the calculated COD:sulphate ratio was < 1. It has been demonstrated that such low ratios can hamper the sulphate reduction process (to $\leq$ 30 % sulphate reduction) and higher COD:sulphate ratios are suggested for achieving high (> 90%) sulphate reduction efficiencies (Villa-Gomez *et al.* 2011; Jing *et al.* 2013).

The sulphate reduction efficiencies achieved with the scourer (95%) and cork (82%) in 47 d of batch operation (Figure 9-2G-H) were comparable to those obtained using the more soluble cheese whey (95%) after ~ 65 d of batch incubation (Martins *et al.* 2009b). In comparison, horse manure, a residue that contains a combination of cellulose (21% *w/w*), lignin (12% *w/w*) and easily available substances (45% *w/w*), supported a sulphate reduction efficiency of 61% after 55 d of batch incubation (Castillo *et al.* 2012). A sulphate reduction efficiency of 77.8% was achieved using rise husk after ~ 17 d of batch incubation (Kijjanapanich *et al.* 2012).

The complexity of L-SRED materials governs the profile of the VFA produced, as evidenced in this study with the scourer and the cork (Figure 9-3). Furthermore, the incubation time influenced the accumulation of short chain VFA, *e.g.* acetate (C_2), propionate (C_3) or higher volatile fatty acids. These VFA are subsequently used as electron donors by heterotrophic SRB (Lissens *et al.* 2004; Bengtsson *et al.* 2008). The SRB present in the inoculum had affinity for butyrate (C_4) and propionate (C_3) (Figure 9-3), demonstrating that the SRB depended on the hydrolysis-fermentation of the supplied L-SRED by the activity of the HFB.

The accumulation of propionate (C_3) suggests the activity of bacteria such as *Clostridia*, which can produce hydrogen from sugars and chemically pre-treated digested vegetables (Goud *et al.* 2012). Propionate (C_3) degrading *Desulfobulbus*-like SRB that consume propionate during the sulphate reduction, using lactate as electron donors, are observed in continuous stirred tank and UASB reactors (Zhao *et al.* 2008; Bertolino *et al.* 2012). Propionate (C_3) can be removed via the succinate pathway to produce acetate (C_2, Lens *et al.* 1998). But propionate (C_3) could also be involved in the production of butyrate (C_4, Lens *et al.* 1996), which is then further oxidized to acetate (C_2) *via* the β-oxidation pathway. Propionate (C_3) can also be involved in the formation of higher VFA like iso-butyrate (C_{4i}) and valerate (C_5) (Lens *et al.* 1996; Lens *et al.* 1998). Alternatively, the iso-butyrate (C_{4i}) and iso-valerate (C_{5i}) could also be the product from the isomerization reactions of their linear structures (Lens *et al.* 1996; Lens *et al.* 1998).

9.4.2 Biological sulphate reduction using L-SRED in an IFBB

Three stages of sulphate reduction and sulphide production were observed during the 49 d of IFBB operation (Figure 9-4C-D), corresponding to the three stages observed in the batch bioreactors using the scourer (see section 9.4.1). According to the literature, sulphate reduction efficiencies might rise up to > 90 % if the COD:sulphate ratio is > 2 (Velasco *et al.* 2008; Villa-Gomez *et al.* 2011). During the early stage of reactor operation (< 10 d), wherein the COD:sulphate ratio was 1.2 (800 mg CODs L^{-1}:682 mg SO_4^{2-} L^{-1}), the sulphate reduction efficiency was as low as 17 ($\pm$ 10)% and sulphide production was not detected (Figure 9-4C-D), presumably little sulphide was formed, which reacted with dissolved metals or lowered the redox potential of the IFBB mixed liquor (Macarie and Guyot 1995). The sulphate reduction efficiency (38 $\pm$ 14%) improved during the second stage (from 10 to 33 d of operation). In the third stage (from 33-49 d), the COD:sulphate ratio was 0.35, the IFBB liquor contained 241 mg CODs L^{-1} and 682 mg SO$_4^{2-}$ L^{-1}, and the sulphate reduction efficiency dropped back to 11 ($\pm$ 8)% (Figure 9-4C).

Based on the hydrodynamic regime imposed by the reactor configuration (either batch or IFBB), the availability of fermented by-products in the IFBB was one of the limiting factors for sulphate reduction using the scourer as L-SRED. Decreasing the influent sulphate concentration (in order to increase the COD:sulphate ratio to > 2) or increasing the HRT to achieve a longer VFA contact time could be a solution to overcome the observed hydrolysis-fermentation drawback of releasing too little CODs for complete reduction of the sulphate present in the synthetic wastewater. In a recent study, the addition of pig farm wastewater treatment sludge to rice husk offered better sulphate reduction efficiencies in continuous bioreactors (Kijjanapanich et al. 2012), because the pig farm wastewater treatment sludge increased the concentration of easily available CODs for the SRB.

Table 9-5 shows SRB were present in the IFBB and this was not limiting the sulphate reduction process in the IFBB with scourer as L-SRED. In other research, the populations of HFB and methanogenic archaea were more pronounced than the SRB population while performing experiments for sulphate reduction using lignocellulosic materials as the substrate and limestone at an HRT of 1 d in a sequencing batch reactor (Pereyra et al. 2010). The pH buffering produced by the limestone could have benefited the methanogenic archaea (Pereyra et al. 2010). In this research, however, the pH depended on the carbonate produced by sulphate reduction mediated by SRB in the IFBB (from 10 to 33 d of IFBB operation, Figure 9-4E), which might thus have hampered the proliferation of an active methanogenic population that scavenged part of the slow release electron donor and thus would reduce the sulphate reduction efficiencies.

9.5   Conclusions

To our knowledge, this is the first time lignocellulosic materials are used as both a carrier material and a SRED in an IFBB without the addition of other easily available or soluble electron donors for SRB under sulphate reducing conditions at 30 °C. Sulphate reduction occurred using a natural scourer (67.7 mg SO$_4^{2-}$ L^{-1} d^{-1} and 0.034 mg SO$_4^{2-}$ mg VSS^{-1} d^{-1}) and cork (12.1 mg SO_4^{2-} L^{-1} d^{-1} and 0.0061 mg SO_4^{2-} mg VSS^{-1} d^{-1}) in batch experiments. The sulphate reduction process in an IFBB using the scourer as L-SRED occurred at an average $V_{r\text{-}IFBB}$ of 167 ($\pm$ 117) mg SO_4^{2-} L^{-1} d^{-1} for 49 d of operation. A $V_{r\text{-}IFBB}$ of 318 ($\pm$ 98) mg SO_4^{2-} L^{-1} d^{-1} was achieved during the best performance (days 10 to 33 of operation). Release of electron donor from the L-SRED, nevertheless, limited and hampered the sulphate reduction process to achieve complete sulphate removal from the synthetic inorganic wastewater supplied to the IFBB.

9.6 References

Alvarado-Lassman A, Rustrián E, García-Alvarado MA, Houbron E (2006) Simultaneous removal of carbon and nitrogen in an anaerobic inverse fluidized bed reactor. Water Sci Technol 54:111–117. doi: 10.2166/wst.2006.493

Angelidaki I, Alves M, Bolzonella D, Borzacconi L, Campos JL, Guwy AJ, Kalyuzhnyi S, Jenicek P, van Lier JB (2009) Defining the biomethane potential (BMP) of solid organic wastes and energy crops: a proposed protocol for batch assays. Water Sci Technol 59:927–934. doi: 10.2166/wst.2009.040

APHA (1999) Standard Methods for the Examination of Water and Wastewater, 20th edn. Washington DC, USA.

Bai H, Kang Y, Quan H, Han Y, Sun J, Feng Y (2013) Treatment of acid mine drainage by sulfate reducing bacteria with iron in bench scale runs. Bioresour Technol 128:818–822. doi: 10.1016/j.biortech.2012.10.070

Bengtsson S, Hallquist J, Werker A, Welander T (2008) Acidogenic fermentation of industrial wastewaters: Effects of chemostat retention time and pH on volatile fatty acids production. Biochem Eng J 40:492–499. doi: 10.1016/j.bej.2008.02.004

Bertolino SM, Rodrigues ICB, Guerra-Sá R, Aquino SF, Leão VA (2012) Implications of volatile fatty acid profile on the metabolic pathway during continuous sulfate reduction. J Environ Manage 103:15–23. doi: 10.1016/j.jenvman.2012.02.022

Borrego J, Carro B, López-González N, de la Rosa J, Grande JA, Gómez T, de la Torre ML (2012) Effect of acid mine drainage on dissolved rare earth elements geochemistry along a fluvial – estuarine system: the Tinto-Odiel Estuary (S.W. Spain). Hydrol Res 43:262–274. doi: 10.2166/nh.2012.012

Cao J, Zhang G, Mao Z-S, Li Y, Fang Z, Yang C (2012) Influence of electron donors on the growth and activity of sulfate-reducing bacteria. Int J Miner Process 106–109:58–64. doi: 10.1016/j.minpro.2012.02.005

Castillo J, Pérez-López R, Sarmiento AM, Nieto JM (2012) Evaluation of organic substrates to enhance the sulfate-reducing activity in phosphogypsum. Sci Total Environ 439:106–113. doi: 10.1016/j.scitotenv.2012.09.035

Celis-García LB, Razo-Flores E, Monroy O (2007) Performance of a down-flow fluidized bed reactor under sulfate reduction conditions using volatile fatty acids as electron donors. Biotechnol Bioeng 97:771–779. doi: 10.1002/bit.21288

Chockalingam E, Sivapriya K, Subramanian S, Chandrasekaran S (2005) Rice husk filtrate as a nutrient medium for the growth of *Desulfotomaculum nigrificans*: Characterisation and sulfate reduction studies. Bioresour Technol 96:1880–1888. doi: 10.1016/j.biortech.2005.01.032

Cohen RRH (2006) Use of microbes for cost reduction of metal removal from metals and mining industry waste streams. J Clean Prod 14:1146–1157. doi: 10.1016/j.jclepro.2004.10.009

Corma A, Iborra S, Velty A (2007) Chemical routes for the transformation of biomass into chemicals. Chem Rev 107:2411–2502. doi: 10.1021/cr050989d

Deng D, Lin L-S (2013) Two-stage combined treatment of acid mine drainage and municipal wastewater. Water Sci Technol 67:1000–1007. doi: 10.2166/wst.2013.653

Equeenuddin SM, Tripathy S, Sahoo PK, Panigrahi MK (2010) Hydrogeochemical characteristics of acid mine drainage and water pollution at Makum Coalfield, India. J Geochemical Explor 105:75–82. doi: 10.1016/j.gexplo.2010.04.006

Foucher S, Battaglia-Brunet F, Ignatiadis I, Morin D (2001) Treatment by sulfate-reducing bacteria of Chessy acid-mine drainage and metals recovery. Chem Eng Sci 56:1639–1645. doi: 10.1016/S0009-2509(00)00392-4

Garcia-Calderon D, Buffiere P, Moletta R, Elmaleh S (1998) Anaerobic digestion of wine distillery wastewater in down-flow fluidized bed. Water Res 32:3593–3600. doi: 10.1016/S0043-1354(98)00134-1

Goud RK, Raghavulu SV, Mohanakrishna G, Naresh K, Mohan SV (2012) Predominance of *Bacilli* and *Clostridia* in microbial community of biohydrogen producing biofilm sustained under diverse acidogenic operating conditions. Int J Hydrogen Energy 37:4068–

4076. doi: 10.1016/j.ijhydene.2011.11.134

Grein F, Ramos AR, Venceslau SS, Pereira IAC (2013) Unifying concepts in anaerobic respiration: insights from dissimilatory sulfur metabolism. Biochim Biophys Acta 1827:145–160. doi: 10.1016/j.bbabio.2012.09.001

Iverson WP (1966) Growth of *Desulfovibrio* on the surface of agar media. Appl Microbiol 14:529–534.

Janyasuthiwong S, Rene ER, Esposito G, Lens PNL (2016) Effect of pH on the performance of sulfate and thiosulfate-fed sulfate reducing inverse fluidized bed reactors. J Environ Eng 142:1–11. doi: 10.1061/(ASCE)EE.1943-7870.0001004

Jiménez-Rodríguez AM, Durán-Barrantes MM, Borja R, Sánchez E, Colmenarejo MF, Raposo F (2009) Heavy metals removal from acid mine drainage water using biogenic hydrogen sulphide and effluent from anaerobic treatment: Effect of pH. J Hazard Mater 165:759–765. doi: 10.1016/j.jhazmat.2008.10.053

Jing Z, Hu Y, Niu Q, Liu Y, Li Y-Y, Wang XC (2013) UASB performance and electron competition between methane-producing archaea and sulfate-reducing bacteria in treating sulfate-rich wastewater containing ethanol and acetate. Bioresour Technol 137:349–357. doi: 10.1016/j.biortech.2013.03.137

Kaksonen AH, Puhakka JA (2007) Sulfate reduction based bioprocesses for the treatment of acid mine drainage and the recovery of metals. Eng Life Sci 7:541–564. doi: 10.1002/elsc.200720216

Kamm B, Kamm M (2004) Principles of biorefineries. Appl Microbiol Biotechnol 64:137–145. doi: 10.1007/s00253-003-1537-7

Kijjanapanich P, Do AT, Annachhatre AP, Esposito G, Yeh DH, Lens PNL (2014) Biological sulfate removal from construction and demolition debris leachate: Effect of bioreactor configuration. J Hazard Mater 269:38–44. doi: 10.1016/j.jhazmat.2013.10.015

Kijjanapanich P, Pakdeerattanamint K, Lens PNL, Annachhatre AP (2012) Organic substrates as electron donors in permeable reactive barriers for removal of heavy metals from acid mine drainage. Environ Technol 33:2635–2644. doi: 10.1080/09593330.2012.673013

Ko J-J, Shimizu Y, Ikeda K, Kim S-K, Park C-H, Matsui S (2009) Biodegradation of high molecular weight lignin under sulfate reducing conditions: Lignin degradability and degradation by-products. Bioresour Technol 100:1622–1627. doi:

10.1016/j.biortech.2008.09.029

Kump LR, Pavlov A, Arthur MA (2005) Massive release of hydrogen sulfide to the surface ocean and atmosphere during intervals of oceanic anoxia. Geology 33:397–400. doi: 10.1130/G21295.1

Lakaniemi AM, Nevatalo LM, Kaksonen AH, Puhakka JA (2010) Mine wastewater treatment using *Phalaris arundinacea* plant material hydrolyzate as substrate for sulfate-reducing bioreactor. Bioresour Technol 101:3931–3939. doi: 10.1016/j.biortech.2010.01.020

Lens P, Vallero M, Esposito G, Zandvoort M (2002) Perspectives of sulfate reducing bioreactors in enviromental biotechnology. Re/Views Environ Sci Bio/Technology 1:311–325.

Lens PN, Dijkema C, Stams AJ (1998) 13C-NMR study of propionate metabolism by sludges from bioreactors treating sulfate and sulfide rich wastewater. Biodegradation 9:179–186. doi: 10.1023/A:1008395724938

Lens PNL, O'flaherty V, Dijkema C, Colleran E, Stams AJM (1996) Propionate degradation by mesophilic anaerobic sludge: Degradation pathways and effects of other volatile fatty acids. J Ferment Bioeng 82:387–391. doi: 10.1016/0922-338X(96)89156-2

Lissens G, Verstraete W, Albrecht T, Brunner G, Creuly C, Seon J, Dussap G, Lasseur C (2004) Advanced anaerobic bioconversion of lignocellulosic waste for bioregenerative life support following thermal water treatment and biodegradation by Fibrobacter succinogenes. Biodegradation 15:173–183.

Lopes SIC, Sulistyawati I, Capela MI, Lens PNL (2007) Low pH (6, 5 and 4) sulfate reduction during the acidification of sucrose under thermophilic (55°C) conditions. Process Biochem 42:580–591. doi: 10.1016/j.procbio.2006.11.004

Macarie H, Guyot JP (1995) Use of ferrous sulphate to reduce the redox potential and allow the start-up of UASB reactors treating slowly biodegradable compounds: application to a wastewater containing 4-Methylbenzoic acid. Environ Technol 16:1185–1192. doi: 10.1080/09593331608616354

Malherbe S, Cloete TE (2002) Lignocellulose biodegradation: fundamentals and applications. Rev Environ Sci Biotechnol 1:105–114. doi: 10.1023/A:1020858910646

Mapanda F, Nyamadzawo G, Nyamangara J, Wuta M (2007) Effects of discharging acid-mine drainage into evaporation ponds lined with clay on chemical quality of the surrounding soil and water. Phys Chem Earth, Parts A/B/C 32:1366–1375. doi: 10.1016/j.pce.2007.07.041

Martins M, Faleiro ML, Barros RJ, Veríssimo AR, Barreiros MA, Costa MC (2009a) Characterization and activity studies of highly heavy metal resistant sulphate-reducing bacteria to be used in acid mine drainage decontamination. J Hazard Mater 166:706–713. doi: 10.1016/j.jhazmat.2008.11.088

Martins M, Faleiro ML, Barros RJ, Veríssimo AR, Costa MC (2009b) Biological sulphate reduction using food industry wastes as carbon sources. Biodegradation 20:559–567. doi: 10.1007/s10532-008-9245-8

Matassa S, Boon N, Verstraete W (2014) Resource recovery from used water: the manufacturing abilities of hydrogen-oxidizing bacteria. Water Res 68:467–478. doi: 10.1016/j.watres.2014.10.028

Monterroso C, Macías F (1998) Drainage waters affected by pyrite oxidation in a coal mine in Galicia (NW Spain): composition and mineral stability. Sci Total Environ 216:121–132. doi: 10.1016/S0048-9697(98)00149-1

NOM-127-SSA1 (1994) Salud ambiental, agua para uso y consumo humano - límites permisibles de calidad y tratamientos a que debe someterse el agua para su potabilización. Mexico (http://www.salud.gob.mx/unidades/cdi/nom/127ssa14.html, last access: 21.04.2017).

Parshina SN, Kijlstra S, Henstra AM, Sipma J, Plugge CM, Stams AJM (2005a) Carbon monoxide conversion by thermophilic sulfate-reducing bacteria in pure culture and in co-culture with *Carboxydothermus hydrogenoformans*. Appl Microbiol Biotechnol 68:390–396. doi: 10.1007/s00253-004-1878-x

Parshina SN, Sipma J, Nakashimada Y, Henstra AM, Smidt H, Lysenko AM, Lens PNL, Lettinga G, Stams AJM (2005b) *Desulfotomaculum carboxydivorans sp. nov.*, a novel sulfate-reducing bacterium capable of growth at 100% CO. Int J Syst Evol Microbiol 55:2159–2165. doi: 10.1099/ijs.0.63780-0

Pereyra LP, Hiibel SR, Prieto Riquelme M V, Reardon KF, Pruden A (2010) Detection and quantification of functional genes of cellulose- degrading, fermentative, and sulfate-reducing bacteria and methanogenic archaea. Appl Environ Microbiol 76:2192–2202. doi: 10.1128/AEM.01285-09

Pérez J, Muñoz-Dorado J, De La Rubia T, Martínez J (2002) Biodegradation and biological treatments of cellulose, hemicellulose and lignin: An overview. Int Microbiol 5:53–63. doi: 10.1007/s10123-002-0062-3

Reyes-Alvarado LC, Okpalanze NN, Rene ER, Rustrian E, Houbron E, Esposito G, Lens PNL (2017) Carbohydrate based polymeric materials as slow release electron donors for sulphate removal from wastewater. J Environ Manage. doi: 10.1016/j.jenvman.2017.05.074

Sarti A, Zaiat M (2011) Anaerobic treatment of sulfate-rich wastewater in an anaerobic sequential batch reactor (AnSBR) using butanol as the carbon source. J Environ Manage 92:1537–1541. doi: 10.1016/j.jenvman.2011.01.009

Sipma J, Meulepas RJW, Parshina SN, Stams AJM, Lettinga G, Lens PNL (2004) Effect of carbon monoxide, hydrogen and sulfate on thermophilic (55 °C) hydrogenogenic carbon monoxide conversion in two anaerobic bioreactor sludges. Appl Microbiol Biotechnol 64:421–428. doi: 10.1007/s00253-003-1430-4

Sipma J, Osuna MB, Parshina SN, Lettinga G, Stams AJM, Lens PNL (2007) H_2 enrichment from synthesis gas by *Desulfotomaculum carboxydivorans* for potential applications in synthesis gas purification and biodesulfurization. Appl Microbiol Biotechnol 76:339–347. doi: 10.1007/s00253-007-1028-3

Tanobe VOA, Sydenstricker THD, Munaro M, Amico SC (2005) A comprehensive characterization of chemically treated Brazilian sponge-gourds (*Luffa cylindrica*). Polym Test 24:474–482. doi: 10.1016/j.polymertesting.2004.12.004

Technical association of the pulp and paper industry (2002) Tappi Test Methods. New York, USA

Teclu D, Tivchev G, Laing M, Wallis M (2009) Determination of the elemental composition of molasses and its suitability as carbon source for growth of sulphate-reducing bacteria. J Hazard Mater 161:1157–1165. doi: 10.1016/j.jhazmat.2008.04.120

Thabet OBD, Bouallagui H, Cayol J, Ollivier B, Fardeau M-L, Hamdi M (2009) Anaerobic degradation of landfill leachate using an upflow anaerobic fixed-bed reactor with microbial sulfate reduction. J Hazard Mater 167:1133–1140. doi: 10.1016/j.jhazmat.2009.01.114

US Environmental Protection Agency (1994) Secondary Drinking Water Standards: Guidance for Nuisance Chemicals, Washington DC, USA.

Velasco A, Ramírez M, Volke-Sepúlveda T, González-Sánchez A, Revah S (2008) Evaluation of feed COD/sulfate ratio as a control criterion for the biological hydrogen sulfide production and lead precipitation. J Hazard Mater 151:407–413. doi: 10.1016/j.jhazmat.2007.06.004

Villa-Gomez D, Ababneh H, Papirio S, Rousseau DPL, Lens PNL (2011) Effect of sulfide concentration on the location of the metal precipitates in inversed fluidized bed reactors. J Hazard Mater 192:200–207. doi: 10.1016/j.jhazmat.2011.05.002

Whiteley CG, Lee D-J (2006) Enzyme technology and biological remediation. Enzyme Microb Technol 38:291–316. doi: 10.1016/j.enzmictec.2005.10.010

WHO (1994) Guidelines for Drinking Water Quality Control, World Health Organization, Geneva, Switzerland.

Zhao Y, Ren N, Wang A (2008) Contributions of fermentative acidogenic bacteria and sulfate-reducing bacteria to lactate degradation and sulfate reduction. Chemosphere 72:233–242. doi: 10.1016/j.chemosphere.2008.01.046

Chapter 10
General discussion and perspectives

10.1 Introduction

The treatment of sulphate rich wastewaters should be done with commitment from the polluting industries as well at the societal/municipal level to avoid the contribution of anthropogenic activities on the natural sulphur cycle. In addition, the optimal use of the natural resources would lead to the sustainability of biotechnological processes and yield economic and environmental benefits to the society (Omer, 2008). In this way, implementing strategies such as optimizing and intensifying biological process will contribute to sustainable wastewater treatment systems (Muga and Mihelcic, 2008). For instance, integrating different physico-chemical and/or biological processes in the wastewater treatment plant will also reduce the cost of the treatment process (Jachuck *et al.*, 1997), contribute to lower the energy consumption and will certainly represent an economical benefit for the society. Processes treating domestic wastewater normally accomplish a reduction in the chemical oxygen demand, reduce almost one third of carbon dioxide emissions and reduce sludge production by 90% when compared to typical treatment processes, wherein multiple steps are usually involved: sulphate reduction, autotrophic denitrification and nitrification as a part of the process integration step (Lu *et al.*, 2012). Another example is the treatment of inorganic wastewaters like acid mine drainage, where two main goals are usually achieved, *viz.*, the biological removal of sulphate and the *in situ* precipitation of recoverable metals using sulphide produced in the same operational unit (Villa-Gomez *et al.*, 2011). Another aspect that has to be taken into account, nowadays, is the increasing population rate and, therefore, the increasing demand for food, goods and services warrants the design of wastewater treatment units to be more flexible, robust and reliant to pollutant load fluctuations (Sekoulov, 2002).

In this PhD thesis, the removal of sulphate was tested under steady and transient-state feeding conditions, in batch, sequencing batch and continuous bioreactor configurations. The outcome of these experiments contributed to optimize the electron donor supply in sulphate reducing bioreactors treating inorganic sulphate rich wastewaters at different COD/SO$_4^{2-}$ ratios. This research work was carried out with the following objectives: (i) to determine the robustness and resilience of biological sulphate reduction (BSR) under steady, transient state and high rate feeding conditions, (ii) to determine the possible influence of high initial sulphate concentrations on the start-up of sequencing batch bioreactors for sulphate reduction, (iii) to elucidate the influence of initial electron donor, NH$_4^+$ and sulphate concentration on BSR, (iv) to study the BSR using carbohydrate based polymers as slow release electron donors, and (v) to study the BSR using lignocellulosic polymers as slow release electron donors.

10.2 General discussion and conclusions

The use of anaerobic bioreactors for the treatment of inorganic sulphate rich wastewaters have been widely reviewed in the literature (Kaksonen and Puhakka, 2007; Lens *et al.*, 2002; Papirio *et al.*, 2013). For instance, BSR has been studied in different bioreactor configurations: the batch reactor (Deng and Lin, 2013), sequencing batch reactor (Torner-Morales and Buitron, 2010), anaerobic biofilm sequencing batch reactor (Sarti *et al.*, 2009), up-flow anaerobic sludge blanket reactor (UASB) (Vallero *et al.*, 2004), the extended granular sludge bed reactor (EGSB) (Dries *et al.*, 1998), fixed bed reactor (Thabet *et al.*, 2009), fluidized bed reactor (Nevatalo *et al.*, 2010) and gas lift reactor (Sipma *et al.*, 2007). In all these reactor configurations, the effect of COD/SO$_4^{2-}$ ratio, the use of different electron donors, mesophilic and thermophilic conditions and the effect of hydraulic residence time (HRT) were tested.

The lack of organic matter or COD in the sulphate rich wastewater is an important factor affecting the economics of the process. As it is well known, the use of pure chemicals as electron donors will only increase the cost of sulphate removal. The utilization of different commercially as well as naturally available low-cost electron donors has been discussed extensively in the literature (Liamleam and Annachhatre, 2007). It is evident that the COD/SO$_4^{2-}$ ratio is an important process variable that affects the performance of BSR in bioreactors. In previous works, a COD/SO$_4^{2-}$ ratio of 0.67 was recommended to completely remove sulphate (100% removal efficiency, RE) from sulphate rich wastewater. However, it was also recommended to use higher COD/SO$_4^{2-}$ ratios, as high as 10, to overcome the accumulation of organic matter and remove it by means of the methanogenic pathways (Hulshoff Pol *et al.*, 1998). In addition, the presence of excess COD or a lack of COD during sulphate rich wastewater treatment can lead to transient operating conditions, *i.e.* in the form of feast and famine conditions. Such COD transient feeding conditions will lead to shifts in the microbial community (Dar *et al.*, 2008; O'Reilly and Colleran, 2006). For example, at a COD/SO$_4^{2-}$ ratio of 2, the propionate consuming sulphate reducing bacteria (SRB) was the predominant bacteria during the sulphate reducing process. On the contrary, acetate and hydrogen consuming methanogenic archaea could not be out competed at this ratio (2). However, at COD/SO$_4^{2-}$ ratios > 10 methanogenic archaea were predominant (Dar *et al.*, 2008; O'Reilly and Colleran, 2006). In the literature, COD/SO$_4^{2-}$ ratios of 2 to 2.5 is generally recommended to achieve sulphate RE in the order of ≥ 90 % (O'Reilly and Colleran, 2006; Torner-Morales and Buitrón, 2009; Velasco *et al.*, 2008).

Given this scenario, in Chapter 3, the robustness and resilience of sulphate reduction to feast and famine (FF) conditions was studied in an inverse fluidized bed bioreactors (IFBB). The robustness was shown when the sulphate RE under transient feeding conditions was very similar to those observed during steady feeding conditions and the RE values ranged between 61 and 71%. Furthermore, the hydrodynamics of the bioreactor also directly influenced the robustness, resilience and adaptation time of the IFBB. Based on the results obtained from the IFBB, the COD/SO$_4^{2-}$ ratio was identified as the most important factor that affected the BSR, followed by the COD loading rate in the IFBB.

Furthermore, in our concentrated efforts to develop versatile bioreactors for BSR, the effect of high rate (HRT $\leq$ 0.25 d) feeding conditions was studied in Chapter 4 using an IFBB. It was observed that the sulphate RE (79%) was not affected despite the low HRT (0.125 d) tested and this RE corresponded to the highest removal rate observed (4,866 mg SO$_4^{2-}$. L^{-1} d^{-1}) at a COD/SO$_4^{2-}$ ratio of 2.3. From that study, it was concluded that the sulphate reduction was majorly limited by the influent COD concentrations and it was not affected by the hydrodynamic conditions. Several studies have also reported the effects of COD limitation and the influence of the COD/SO$_4^{2-}$ ratio on the sulphate removal in IFBB (Celis-García et al., 2007; Papirio et al., 2013; Villa-Gomez et al., 2011). When the experimental data was fitted to the Grau second order model for substrate removal (Grau et al., 1975), the model fitted the high rate removal performance with an $r^2 > 0.96$. This clearly implies that the sulphate reduction was not affected by the low HRT (0.125) examined in the IFBB. The Grau second order model of substrate removal takes into account the substrate:biomass ratio for estimating the kinetic constants. This clearly demonstrates the fact that biomass, i.e. the volatile suspended solids (VSS), was not washed out from the reactor at the lowest HRT (0.125 d) tested and, therefore, the VSS concentration was not limiting the BSR in the IFBB.

Apart from the COD/SO$_4^{2-}$ ratio, the nature of the electron donor also affects the microbial population during the sulphate removal process. According to Zhao et al. (2010), when simple carbon sources are fed to the bioreactor, they are readily utilized by the SRB and this increases the fraction of SRB, improves sulphate reduction and reduces the start-up time. In that study, the authors also demonstrated that lactate benefits the proliferation of δ-proteobacteria rather than firmicutes. In Chapter 5, the effect of initial sulphate concentration on the start-up phase of BSR was studied at a constant COD/SO$_4^{2-}$ ratio (2.4) in two sequencing batch reactors (SBR) operated in parallel. The SBR using 2.5 g SO$_4^{2-}$. L^{-1} showed lower resistance to remove sulphate, reaching RE of 62 ($\pm$ 25)% from 0 to 8 d of operation. In comparison, the SBR using

0.4 g SO_4^{2-}. L^{-1} showed a sulphate RE of 22 ($\pm$ 15)% from 0 to 12 d of operation. Using the mixed anaerobic consortium in the SBR, the sulphate reduction process was promoted during the start-up phase. A similar observation was also reported in the literature, wherein the specific growth rate of a pure culture of SRB was optimally promoted at a sulphate concentration of 2.5 g $SO_4^{2-}.L^{-1}$ (Al Zuhair et al., 2008). Additionally, the adaptation time of the BSR also influenced the accumulation or consumption of either propionate or acetate during the start-up phase.

In Chapter 6, the influence of the initial COD, NH_4^+ and sulphate concentrations on the BSR was ascertained in batch bioreactors. The main results of this study showed that NH_4^+ had very little effect on the sulphate removal rates while a major NH_4^+ influence was observed on the electron donor uptake during sulphate reduction. In addition, the electron donor utilization via the BSR process improved simultaneously to the decreasing initial electron donor concentrations. Presumably, this behaviour is related to the prevalence or dominance of certain type of bacteria under decreased initial COD/SO_4^{2-} ratio. According to Dar et al. (2008), the incomplete lactate oxidizing SRB were predominant at COD/SO_4^{2-} ratios < 0.4. In another report, *Desulfovibrio desulfuricans* decreased the electron donor uptake while increasing the NH_4^+ famine conditions during BSR, and due to the famine conditions, the cell size was negatively affected and the cell carbon content decreased (Okabe et al., 1992). For instance, according to Habicht et al. (2005), during sulphate reduction experiments with the *Archaeoglobus fulgidus* strain Z and under sulphate feast conditions, the biomass production and sulphate consumption rates were higher compared to sulphate starving conditions. Besides, according to the authors, the carbon uptake for biomass production rate was also greater during sulphate feast conditions.

The BSR process also proved to be robust when its performance was evaluated in SBRs that were operated at a constant COD/SO_4^{2-} ratio of 2.4 (Chapter 7). Six SBR were tested in this part of the research. The SBR R1L showed a low RE of 79 ($\pm$ 17)% at low sulphate concentrations (0.4 g $SO_4^{2-}.L^{-1}$) and under steady feeding conditions. On the other hand, the robustness of BSR was shown by means of the similarities on sulphate RE noticed during feast (92%) and famine (86-90 %) conditions in the SBR R2L and R3L, respectively. In comparison, the control reactor R1H showed a RE of 94 ($\pm$ 8)% at a sulphate concentration of 2.5 g SO_4^{2-} .L^{-1}. When SBR R2H and SBR R3H were subjected to harsher operating conditions, i.e. sulphate concentration of 15 g $SO_4^{2-}.L^{-1}$, the sulphate reduction process crashed abruptly due to exceedingly high sulphate concentrations and inhibition of the biomass.

Therefore, the BSR process is more affected by: (i) the concentration of SRB in the inoculum, *i.e.* the SO_4^{2-}:SRB ratio and (ii) the transient conditions of macronutrients (C, N, S) that are used to build new biomass (*e.g.* proteins, DNA, exopolysaccharides, cell wall). Thus, continuous bioreactors operating at high rate feeding conditions (HRT < 0.125 d) are more susceptible to loss of active biomass followed by a complete reactor failure during sulphate reduction than reactors with lower or negligible biomass losses, as in the case of SBR or batch bioreactors.

This PhD work also focused on the use of slow release electron donors (SRED) for BSR in batch and IFBB. The SRED are organic polymers that cannot be directly used by the SRB, but rather they can be hydrolyzed first by hydrolytic fermentative bacteria and later the SRB can use the hydrolysis and/or the fermentation products to reduce the sulphate present in wastewater. BSR experiments carried out using carbohydrate based polymers (CBP) as slow release electron donors (SRED) (Chapter 8), the sulphate reduction rates were affected by the hydrolysis-fermentation rates of the CBP as well as the nature of the CBP. The high sulphate RE using cellulose (filter paper > 98 %) suggested the utilization of polymers with higher degree of polymerization as lignocellulose (L) to be tested as SRED. The use of scourer and lignin as L-SRED yielded a sulphate RE > 82 % (Chapter 9). The experiments in batch and IFBB, by means of their respective incubation time and relative short HRT (1 d), confirmed that the L-SRED hydrolysis-fermentation was the rate limiting step during the BSR.

From a pollution prevention and resource recovery view-point, *in situ* metal precipitation with biogenic sulphide is an interesting growing field of research (Lewis, 2010). In bioreactors, the settleability and the quality of the metal sulphide crystals is influenced by the sulphide concentration produced by SRB (Villa-Gomez *et al.*, 2012). The use of SRED offers to control the production of sulphide in order to avoid supersaturation and thus generate the desired optimal sulphide concentration required for selective metal precipitation (Sampaio *et al.*, 2009).

10.3 Future research work

The BSR and the sulphate rich wastewater treatment process offers many other aspects of study for process intensification and optimization. Future research work concerning sulphate reduction should be aimed at investigating the following aspects in bioreactors:

The IFBB is capable of maintaining the active sulphate reducing biomass and this was clearly explained by the Grau second order model of substrate removal under the conditions tested in this research. From a practical perspective, the bioreactors hydrodynamics should be tested under more stressful or harsh operating conditions in order to optimize BSR. The effect of IFBB design parameters such as reactor geometry, height:diameter (H/D) ratio, different support materials, liquid recirculation velocity and gas/liquid hold up should be varied and its effect on BSR should be ascertained during long-term operations.

By employing extensive bio-molecular level investigations, studies should be undertaken to monitor the evolution of microbial communities and characterize the microbial population dynamics at different HRT and under adequate supply of the carbon and sulphate source.

Furthermore, it will be useful to investigate the impact of sulphate loading rate on the SRB physiology and ascertain how the SRB manages the stressful environment. Studies should be aimed towards understanding the impact on biochemical capacity, cellular survival capacity and the role of different environmental stimuli on individual microorganisms. It is worthwhile to investigate the self-detoxification mechanism involved during the BSR process and identify the possible scenarios that stimulates the expression of known stress-response genes.

Moreover, due to the increasing demand of bioplastics in the world market, the production of biopolymers is an interesting field of research. SRB has the capacity to produce polyhydroxyalkanoates (PHA), but not under lactate utilization (Hai *et al.*, 2004). However, the production of poly-thioesters (Lütke-Eversloh and Steinbüchel, 2004) could be investigated by means of utilizing the sulphur present in the sulphate rich wastewater.

10.4 References

Al Zuhair, S., H El-Naas, M., Al Hassani, H., 2008. Sulfate inhibition effect on sulfate reducing bacteria. J. Biochem. Technol. 1, 39-44.

Celis-García, L.B., Razo-Flores, E., Monroy, O., 2007. Performance of a down-flow fluidized bed reactor under sulfate reduction conditions using volatile fatty acids as electron donors. Biotechnol. Bioeng. 97, 771-779.

Dar, S.A, Kleerebezem, R., Stams, A.J.M., Kuenen, J.G., Muyzer, G., 2008. Competition and coexistence of sulfate-reducing bacteria, acetogens and methanogens in a lab-scale anaerobic bioreactor as affected by changing substrate to sulfate ratio. Appl. Microbiol. Biotechnol. 78, 1045-1055.

Deng, D., Lin, L.-S., 2013. Two-stage combined treatment of acid mine drainage and municipal wastewater. Water Sci. Technol. 67, 1000-1007.

Dries, J., De Smul, A., Goethals, L., Grootaerd, H., Verstraete, W., 1998. High rate biological treatment of sulfate-rich wastewater in an acetate-fed EGSB reactor. Biodegradation 9, 103-111.

Grau, P., Dohányos, M., Chudoba, J., 1975. Kinetics of multicomponent substrate removal by activated sludge. Water Res. 9, 637-642.

Habicht, K.S., Salling, L., Thamdrup, B., Canfield, D.E., 2005. Effect of low sulfate concentrations on lactate oxidation and isotope fractionation during sulfate reduction by *Archaeoglobus fulgidus* strain Z. Appl. Environ. Microbiol. 71, 3770-3777.

Hai, T., Lange, D., Rabus, R., Steinbuchel, A., 2004. Polyhydroxyalkanoate (PHA) accumulation in sulfate-reducing bacteria and identification of a Class III PHA synthase (PhaEC) in *Desulfococcus multivorans*. Appl. Environ. Microbiol. 70, 4440-4448.

Hulshoff Pol, L.W., Lens, P.N.L., Stams, A.J.M., Lettinga, G., 1998. Anaerobic treatment of sulphate-rich wastewaters. Biodegradation 9, 213-224.

Jachuck, R., Lee, J., Kolokotsa, D., Ramshaw, C., Valachis, P., Yanniotis, S., 1997. Process intensification for energy saving. Appl. Therm. Eng. 17, 861-867.

Kaksonen, A.H., Puhakka, J.A., 2007. Sulfate reduction based bioprocesses for the treatment of acid mine drainage and the recovery of metals. Eng. Life Sci. 7, 541-564.

Lens, P., Vallerol, M., Esposito, G., Zandvoort, M., 2002. Perspectives of sulfate reducing bioreactors in environmental biotechnology. Rev. Environ. Sci. Bio/Technology 1, 311-325.

Lewis, A.E., 2010. Review of metal sulphide precipitation. Hydrometallurgy 104, 222-234.

Liamleam, W., Annachhatre, A.P., 2007. Electron donors for biological sulfate reduction. Biotechnol. Adv. 25, 452-463.

Lu, H., Wu, D., Jiang, F., Ekama, G. a, van Loosdrecht, M.C.M., Chen, G.-H., 2012. The demonstration of a novel sulfur cycle-based wastewater treatment process: Sulfate reduction, autotrophic denitrification, and nitrification integrated (SANI®) biological nitrogen removal process. Biotechnol. Bioeng. 109, 2778-2789.

Lütke-Eversloh, T., Steinbüchel, A., 2004. Microbial polythioesters. Macromol. Biosci. 4, 165-174.

Muga, H.E., Mihelcic, J.R., 2008. Sustainability of wastewater treatment technologies. J. Environ. Manage. 88, 437-447.

Nevatalo, L.M., Mäkinen, A.E., Kaksonen, A.H., Puhakka, J.A, 2010. Biological hydrogen sulfide production in an ethanol-lactate fed fluidized-bed bioreactor. Bioresour. Technol. 101, 276-284.

O'Reilly, C., Colleran, E., 2006. Effect of influent COD/SO_4^{2-} ratios on mesophilic anaerobic reactor biomass populations: physico-chemical and microbiological properties. FEMS Microbiol. Ecol. 56, 141-153.

Okabe, S., Nielsen, P.H., Charcklis, W.G., 1992. Factors affecting microbial sulfate reduction by *Desulfovibrio desulfuricans* in continuous culture: limiting nutrients and sulfide concentration. Biotechnol. Bioeng. 40, 725-734.

Omer, A.M., 2008. Green energies and the environment. Renew. Sustain. Energy Rev. 12, 1789-1821.

Papirio, S., Esposito, G., Pirozzi, F., 2013. Biological inverse fluidized-bed reactors for the treatment of low pH- and sulphate-containing wastewaters under different COD/SO_4^{2-} conditions. Environ. Technol. 34, 1141-1149.

Papirio, S., Villa-Gomez, D.K., Esposito, G., Pirozzi, F., Lens, P.N.L., 2013. Acid mine drainage treatment in fluidized-bed bioreactors by sulfate-reducing bacteria: a critical review. Crit. Rev. Environ. Sci. Technol. 43, 2545-2580.

Sampaio, R.M.M., Timmers, R. a, Xu, Y., Keesman, K.J., Lens, P.N.L., 2009. Selective precipitation of Cu from Zn in a pS controlled continuously stirred tank reactor. J. Hazard. Mater. 165, 256-265.

Sarti, A., Silva, A.J., Zaiat, M., Foresti, E., 2009. The treatment of sulfate-rich wastewater using an anaerobic sequencing batch biofilm pilot-scale reactor. Desalination 249, 241-246.

Sekoulov, I., 2002. Sustainable development of wastewater treatment strategies for the food industries. Water Sci. Technol. 45, 315-320.

Sipma, J., Osuna, M.B., Lettinga, G., Stams, A.J.M., Lens, P.N.L., 2007. Effect of hydraulic retention time on sulfate reduction in a carbon monoxide fed thermophilic gas lift reactor. Water Res. 41, 1995-2003.

Thabet, O.B.D., Bouallagui, H., Cayol, J., Ollivier, B., Fardeau, M.-L., Hamdi, M., 2009. Anaerobic degradation of landfill leachate using an upflow anaerobic fixed-bed reactor with microbial sulfate reduction. J. Hazard. Mater. 167, 1133-1140.

Torner-Morales, F.J., Buitron, G., 2010. Kinetic characterization and modeling simplification of an anaerobic sulfate reducing batch process. J. Chem. Technol. Biotechnol. 85, 453-459.

Torner-Morales, F.J., Buitrón, G., 2009. Kinetic characterization and modeling simplification of an anaerobic sulfate reducing batch process. J. Chem. Technol. Biotechnol. 85, 453-459.

Vallero, M.V.G., Sipma, J., Lettinga, G., Lens, P.N.L., 2004. High-rate sulfate reduction at high salinity (up to 90 mS.cm^{-1}) in mesophilic UASB reactors. Biotechnol. Bioeng. 86, 226-235.

Velasco, A., Ramírez, M., Volke-Sepúlveda, T., González-Sánchez, A., Revah, S., 2008. Evaluation of feed COD/sulfate ratio as a control criterion for the biological hydrogen sulfide production and lead precipitation. J. Hazard. Mater. 151, 407-413.

Villa-Gomez, D., Ababneh, H., Papirio, S., Rousseau, D.P.L., Lens, P.N.L., 2011. Effect of sulfide concentration on the location of the metal precipitates in inversed fluidized bed reactors. J. Hazard. Mater. 192, 200-207.

Villa-Gomez, D.K., Papirio, S., van Hullebusch, E.D., Farges, F., Nikitenko, S., Kramer, H., Lens, P.N.L., 2012. Influence of sulfide concentration and macronutrients on the characteristics of metal precipitates relevant to metal recovery in bioreactors. Bioresour. Technol. 110, 26-34.

Zhao, Y.-G., Wang, A.-J., Ren, N.-Q., 2010. Effect of carbon sources on sulfidogenic bacterial communities during the starting-up of acidogenic sulfate-reducing bioreactors. Bioresour. Technol. 101, 2952-2959.

Biography

Luis Carlos Reyes-Alvarado (born in Córdoba, Veracruz, Mexico) obtained his PhD in Environmental Technology. During the high school, he was trained as a food process technician at CETis 143 in Fortin, Veracruz (Mexico). At this time, Carlos developed a strong interest in the engineering of biological processes. Thus, he joined the Universidad Veracruzana (Mexico) where he obtained the degree of Chemical Engineering and further a Master in Food Science and Technology, the latter with a scholarship provided by CONACyT (National Counsil of Science and Technology). Subsequently, the European Commission awarded him an ALFA grant within the SUPPORT (Sustainable Use of Photosynthesis Products & Optimum Resource Transformation) project at the TU Graz (Austria). Until now, he got trained in processes concerning the production of biofuels by means of pyrolysis and fermentation, both using food wastes. Later, he developed and defended his PhD thesis with auspicious of the European Commission through the Erasmus Mundus Joint Doctorate Programme in Environmental Technologies for Contaminated Solids, Soils and Sediments (ETeCoS[3]) on December 16th, 2016. His research was focused on the optimization of electron donor supply to sulphate reducing bioreactors treating inorganic wastewater rich in sulphate and carried out at different institutions: the UNESCO-IHE (Delft, The Netherlands), the Universidad Veracruzana (Veracruz, Mexico), the INRA-Laboratoire de Biotechnologie de l'Environnement (Narbonne, France) and the University of Cassino and Southern Lazio (Cassino, Italy). L. C. Reyes-Alvarado is interested in understanding the engineering aspects of biological processes, resource recovery from waste and the development of eco-technologies for waste remediation.

Publications

Reyes-Alvarado L.C., Okpalanze N.N., Kankanala D., Rene E.R., Esposito G., Lens P.N.L., (2017) Forecasting the effect of feast and famine conditions on biological sulphate reduction in an anaerobic inverse fluidized bed reactor using artificial neural networks, Process Biochem. 55: 146-161. doi:10.1016/j.procbio.2017.01.021.

Reyes-Alvarado L.C., Okpalanze N.N., Rene E.R., Rustrian E., Houbron E., Esposito G., Lens P.N.L., (2017) Carbohydrate based polymeric materials as slow release electron donors for sulphate removal from wastewater. J Environ Manage. 200: 407-415. doi: 10.1016/j.jenvman.2017.05.074

Reyes-Alvarado L.C., Camarillo-Gamboa A., Rustrian E., Rene E.R., Esposito G., Lens P.N.L., Houbron E., (2017) Lignocellulosic biowastes as carrier material and slow release electron donor for sulphidogenesis of wastewater in an inverse fluidized bed bioreactor. Environmental Science and Pollution Research. 1-14. doi: 10.1007/s11356-017-9334-5

Reyes-Alvarado L.C., Rene E.R., Esposito G., and Lens P.N.L., (2018). Bioprocesses for sulphate removal from wastewater. In: Varjani S., Gnansounou E., Gurunathan B., Pant D., Zakaria Z. (eds.) Waste Bioremediation. Energy, Environment and Sustainability. Springer Nature Singapore Pte Ltd. doi: 10.1007/978-981-10-7413-4_3

Conferences

Luis C. Reyes-Alvarado , Gerardo Ramirez-Morales, Álvaro Camarillo-Gamboa, Eldon R. Rene, Giovanni Esposito, Piet N.L. Lens, Elena Rustrian, Eric Houbron, (2015). Lignocellulosic biowastes as carrier material and slow release electron donors for sulphate removal from wastewater in a continuous inverse fluidized bed bioreactor. 4th IWA Mexico Young Water Professionals Conference. Guanajuato, Gto. México. April 27-29.

Luis C. Reyes-Alvarado, Eldon R. Rene, Gaëlle Santa-Catalina, Frédéric Habouzit, Renau Escudie, Giovanni Esposito, Piet N. L. Lens, Nicolas Bernet, (2016). Impact of initial sulphate concentration on the microbial community and performance of sequencing batch reactors (poster presentation). The 9th CESE Conference, International Conference on Challenges in Environmental Science & Engineering. Kaohsiung, Taiwan. November 6-10.

L. C. Reyes-Alvarado, E. R. Rene, E. Rustrian, G. Esposito, A. Hatzikioseyian, E. Houbron and P. N. L. Lens, (2017). High rate biological sulphate reduction in an inverse fluidized bed reactor at a hydraulic retention time of 3 h. 5th IWA Mexico Young Water Professionals Conference. Morelia, México. May 24-26.

Netherlands Research School for the
Socio-Economic and Natural Sciences of the Environment

D I P L O M A

For specialised PhD training

The Netherlands Research School for the
Socio-Economic and Natural Sciences of the Environment
(SENSE) declares that

Luis Carlos Reyes Alvarado

born on 16 November 1983 in Cordoba, Mexico

has successfully fulfilled all requirements of the
Educational Programme of SENSE.

Cassino, Italy, 16 December 2016

the Chairman of the SENSE board

Prof. dr. Huub Rijnaarts

the SENSE Director of Education

Dr. Ad van Dommelen

The SENSE Research School has been accredited by the Royal Netherlands Academy of Arts and Sciences (KNAW)

K O N I N K L I J K E   N E D E R L A N D S E
A K A D E M I E   V A N   W E T E N S C H A P P E N

The SENSE Research School declares that Mr Luis Reyes Alvarado has successfully fulfilled all
requirements of the Educational PhD Programme of SENSE with a
work load of 47.5 EC, including the following activities:

<u>SENSE PhD Courses</u>

o   Environmental research in context (2013)
o   Research in context activity: 'Contributing to preparation and lecturing for courses on
    environmental bioprocesses and selected topics on environmental biotechnology, Mexico'
    (2014)

<u>Other PhD and Advanced MSc Courses</u>

o   Anaerobic Wastewater Treatment, UNESCO-IHE (2013)
o   ETeCoS3 summer school on contaminated sediments characterization and remediation,
    UNESCO-IHE (2013)
o   ETeCoS[3] summer school on biological treatment of solid waste, Università degli studi di
    Cassino e del Lazio Meridionale (2014)
o   Introductory ETeCoS[3] course Environmental Technology, Università degli studi di Cassino e
    del Lazio Meridionale (2014)
o   P-Rex summer school on implementation of phosphorus recovery from wastewater –why
    and how?, University of Applied Sciences Northwestern Switzerland, (2014)
o   ETeCoS[3] summer school on Contaminated Soils, University Paris-Est (2015)

ETeCoS[3] = Environmental Technologies for Contaminated Solids, Soils and Sediments programme

<u>Management and Didactic Skills Training</u>

o   Supervising three MSc students
o   Assisting at ETeCoS[3] summer school on contaminated sediments characterization and
    remediation (2013)
o   Assisting at G16 2013: 3rd International Conference on Research Frontiers in Chalcogen Cycle
    Science & Technology, Delft, The Netherlands (2013)

<u>Oral Presentations</u>

o   *Optimization of e-donor for SRB: application in acid mine drainage* treatment. Annual
    ETeCoS[3] meeting, 17-18 June 2013, Delft, the Netherlands
o   *Optimization of electron donor dosing for sulphate reducing bacteria in acid mine drainage
    treatment.* Annual ETeCoS[3] meeting, 30 June - 1 July 2014, Cassino, Italy
o   *Optimization of electron donor dosing for sulphate reducing bacteria in sulphate rich
    wastewater treatment.* Annual ETeCoS[3] meeting, 29-30 June 2015, Paris, France
o   Lignocellulosic biowastes as carrier material and slow release electron donors for sulphate
    removal from wastewater in a continuous inverse fluidized bed bioreactor.
    4th IWA Mexico Young Water Professional Conference, 27-29 April 2015, Guanajuato, Mexico

SENSE Coordinator PhD Education

Dr. ing. Monique Gulickx

The main objective of this research was to optimize the electron donor supply in sulphate reducing bioreactors treating sulphate rich wastewater. Two types of electron donor were tested: lactate and slow release electron donors (SRED) such as carbohydrate based polymers (CBP) and lignocellulosic biowastes (L). Biological sulphate reduction (BSR) was evaluated in different bioreactor configurations, namely, the inverse fluidized bed bioreactor (IFBB), sequencing batch reactor (SBR) and batch reactors. The reactors were tested under steady-state, high-rate and transient-state feeding conditions of electron donor and acceptor, respectively. The results showed that the IFBB configuration is robust and resilient to transient and high rate feeding conditions at a hydraulic retention time (HRT) as low as 0.125 d. The BSR was limited by the COD:sulphate ratio (< 1.7). The results from artificial neural network (ANN) modelling showed that the influent sulphate concentrations synergistically affected the COD removal efficiency and the sulphide production. Concerning the role of electron donors, the SRED allowed a BSR > 82% either using CBP or L, in batch bioreactors. The BSR was limited by the hydrolysis-fermentation rate and by the complexity of the SRED.